Phillip Jacklin

Firm & comprehensive

Good solid overview

D0971531

Essay on Atomism

Books by Lancelot Law Whyte

The Atomic Problem: A Challenge to Physicists and Mathematicians (1961)
The Unconscious Before Freud (1960; paper 1962)
Accent on Form (1951)
The Unitary Principle in Physics and Biology (1949)
Everyman Looks Forward (1948)
The Next Development in Man (1948; paper, 1950, 1961)
Critique of Physics (1931)
Archimedes: or, The Future of Physics (1928)

Edited by Mr. Whyte

R. J. Boscovich (1711–1787), Essays on His Life and Work (1961)
Aspects of Form (1951; paper 1961)

Essay on Atomism:

FROM DEMOCRITUS TO 1960

By LANCELOT LAW WHYTE

Wesleyan University Press : MIDDLETOWN, CONNECTICUT

The whole of Nature is of two things built, Atoms and Void.
—Lucretius (c. 60 B.C.)

It is as easy to count atomies, as to resolve the propositions of a lover.
—Shakespeare (1599)

Atoms or Systems into ruin hurled,
And now a bubble burst, and now a world.
—Pope (1734)

And shall an atom of this atom world
Mutter the theme of heaven?
—Young (1742)

It is not strictly demonstrated that atoms are indivisible; but it appears that they are not divided by the laws of nature.
—Voltaire (1756)

Contents

Essay on Atomism

To the Reader

THE conception of atomism has been the spearhead of the advance of science. Atomic ideas have led to the highest adaptive precision which the human brain has yet achieved.

The history of atomism should therefore be of interest to all concerned with the human mind. Whatever the limitations and dangers of atomism, its achievements are unique and worthy of study.

But there is more to it than that. The fertility of the Greek atomic philosophy proves the power of speculative reason. Even Francis Bacon, who had no idea of what was to come, admitted that what he regarded as "undisciplined thinking" had here achieved something extraordinary: an idea so powerful that it had survived centuries of criticism and neglect. Atomism has proved the power of the intellectual imagination to identify aspects of an objective truth deeply rooted in the nature of things. Hidden in the history of atomism and in the abstract mysteries of quantum mechanics there must be, still concealed, a trustworthy foundation for the human intellect.

Is this not a miracle and a promise as worthy of attention as the din of technology or the consolations of religion? A hundred generations ago a few men, trusting in their own minds, pondered on the character of the universe they knew, and produced a new idea, a development of what had gone before but none the less new. This idea not only survived long disuse but, more remarkable, an avalanche of research of a kind inconceivable to the Greeks. And now we are *surprised* that the Greek idea needs modification, that particles are not all permanent. Our astonishment is a tribute to our profound, though only half confessed, belief in the power of ideas. There is majesty in the human mind, and in its history. Its like will not be found on the moon.

Moreover, the story of atomism has now reached a peculiarly interesting stage. During the last thirty years experiment has entered a high-energy realm where fresh ideas appear necessary. But theory has not kept up, and no one knows when and where the new insights will arise. It would be a pity to be alive at such a dramatic time and not to be aware of it. For issues are involved which the layman can appreciate.

It is widely believed that only those who can master the latest quantum mathematics can understand anything of what is happening. That is not so, provided one takes a long view, and that means a historical view, for no one can see far ahead. Against a historical background, the layman can understand what is involved, for example, in the fascinating challenge of continuity and discontinuity expressed in the antithesis of field and particle.

Indeed I shall suggest that a clue to the future must lie in the past, surprising as that may seem. Also that every scientist, and everyone with intellectual curiosity, can learn something useful from a brief study of the history of atomism.

Let me give an example, not taken from the past but from today. We hear of *un*stable particles in physics and of *un*conscious mind in psychology. Is this a mere chance, or a sign of a parallel between the two sciences? Is there some common factor which leads both to name a basic idea in this backhanded manner? I believe that there is and that it throws light on the position of both sciences.

Physics and psychology are each using the negative prefix *un-* to announce a transformation of ideas which is still incomplete. Because they have not yet reached clarity about the new ideas which are necessary, they can so far describe the change only by the denial of an old idea.

More precisely, what has happened is that these sciences are rejecting two ancient views about the nature of things: the Democritan view that *stable* atoms are the basis of phenomena; and the equally old view, sharpened by Descartes, that *conscious* mind is a second, independent mode of existence. These are being discarded because they no longer fit the facts, or do not seem to.

It is somewhat casual to use this brusque little *un-* to overthrow two time-honoured traditions, as has been done, without making clear what changed meanings are hereby given to the old terms *particle* and *mind*. In the mid-XXth century we flatter ourselves on our high self-awareness and sophistication, and great schools are pursuing scientific accuracy, logical precision, linguistic analysis, and the arts of communication. Yet this cultural two-letter bomb is allowed to get away with its dirty work at the crossroads where an old method and a new one intersect.

It cries to heaven: no psychologist has yet provided an acceptable definition of "mind" or "mental" that reveals the character of "unconscious mental processes," and no physicist a lucid definition of "elementary particles" that shows how they can appear and disappear, and why there are so many. This is not the fault of the specialists concerned; no one yet knows enough about either the organ of thought or the organisation of the particles to do this, so that they are bound to cover their ignorance in that prefix. Did ever in the history of the intellect so little conceal so much?

To change the metaphor, physics and psychology are going somewhere, but *where* they do not yet know. Sciences rarely do. But the ticket *un-* at least shows where they are travelling *from:* away from Democritan permanent particles and the Cartesian mind necessarily aware.

If that exhausted the analogy, it would only imply that two scientific revolutions were simultaneously in progress, which would hardly be surprising in a time of intense research. But if we examine their tickets more closely we find, hidden in small print on the back, that they are both travelling away from the same point of origin and in the same general direction: *from* the isolation of supposedly permanent "substances" *towards* the identification of changing relations potentially affecting everything; briefly, *from substances to changing structures of relations.* At least so it seems to an observer himself in the movement.

Up to this point the analogy is precise and reliable. In each case what used to be regarded as an independent permanent substance or mode of existence—the material particle or the conscious mind—has been discovered not to be sufficiently unchanging to be treated as a thing existing in isolation in its own right, but more often to be the opposite: a changing system in a changing environment. The attempted isolation of factors with unchanging properties has apparently failed both in physics and in psychology. The twenty or thirty "elementary particles" of deep physics are no longer the permanent *res extensa* (space-occupying substance), nor is the extended "mind" of deep psychology the *res cogitans* (conscious substance) of the Cartesian doctrine.

I am not suggesting that the analogy goes further: for example, that the unstable particles of nuclear physics are the same as the irrational factors of psycho-pathology, or that in releasing them the reactor does the same as the couch. We must stop the analogy at the recognition that both sciences are now concerned with patterns of changing relations which neither yet knows how to reduce to order. The new feature has two aspects: both a wider context and a finer pattern of changing factors than either Newton or Descartes imagined have now to be taken into account. That is what the analogy can teach us, if we didn't know it before. And even if we did, it calls our attention again to a pervasive feature of the present situation of the sciences. Thus the analogy with psychology may encourage the physicist to take this feature seriously, and to recognize that particles that change are functions of their environment.

If so deep a transformation is in progress, it is not surprising that particle physicists are reaching the conclusion that a new physical language is needed. This is surely right. But it is also widely assumed that this new language must be highly abstract and associated with a calculus so intricate that even its general significance must lie beyond the comprehension of non-specialists. Einstein held that physics necessarily advances from traditional concrete ideas towards more and more abstract ones possessing only highly indirect relations to sense experience, and so far this view has proved correct.

But if this continues, the consequences will be grave. For the entire tradition must then inevitably pass more and more into the hands of a small group of high specialists. Physical power will lie with those communities who teach a selection of their children the necessary super-calculus at the earliest possible age. (The U.S.S.R. and the U.S.A. are already trying to teach ten- to fifteen-year-olds physics via quantum, not classical, mechanics.) Indeed some believe we have already entered the age of Abstract Man.

I am convinced that this is wrong. Nothing is more surprising than the surprises of history, and nothing more untrustworthy than the uncritical extrapolation of the tendencies of the recent past. There are good reasons to expect, within the coming years or decades, a return to a concreteness of basic ideas, to simpler fundamentals easily understood, to principles that will bring exact science closer to the human person. The sciences are manifestly converging; atom, man, and universe are, in an objective and inescapable sense, deeply related. Moreover—and this is the primary reason for my view—physics is ultimately concerned, not with features in some abstract higher-dimensional manifold, but with observable events in the three-dimensional space in which man spends his life. However far physics may temporarily and for special purposes reach out into abstraction, it must sooner or later return to man's actions and thoughts in this world of processes in three-dimensional space, in order to describe phenomena.

Indeed the graver the "crisis" in physics is taken to be, the more likely it is that the next fertile step will be towards a more immediate and concrete foundation, rather than to further abstractions. For the most productive novelties often spring, in thought as in biological evolution, from more primitive and simpler forms, rather than from differentiated ones which, through their elaboration, have become too specialized to be adaptable to new tasks.

If a future theory is to unify atomic knowledge, it must contain all established theories as particular cases, and the path towards its more general principles must lie through the discarding of redundant restrictions. Somewhere, hidden in the unconscious assumptions of the physics of 1960, there must lie redundant traditional elements which can be dropped in creating a more powerful future theory. Thus the major advances of theoretical physics, once it has reached the stage of a comprehensive system, must arise from the rendering conscious and the elimination of unnecessary unconscious features which have either crept into, or always been present in, traditional thought. Such systematic errors of theory can seldom be discovered by direct attack; it is easier to uncover them by studying how and why physical theory took the path it did. That is why a clue to the future can sometimes be found in the past, and this is my reason for exploring the history of atomism.

But that is not why I believe it worth while publishing a brief sketch and interpretation of the history of atomic ideas.

Until a few years ago I imagined that courses in physics included an outline of the history of atomism. Unfortunately that is not so. Many universities, teachers, and students consider they have no time for that.

When Rutherford took over the Cavendish Laboratory he was shocked to discover that a younger generation thought that science began in their day. I have been disturbed to realize that every scientific generation, measured by its most vocal members, exaggerates the historical importance of its own heroes. The character of science as continual advance renders this bias inevitable, for there is a perpetual temptation to study the latest and to neglect the past. The story of atomism is sufficiently covered if one knows something about Democritus and Dalton.

The following blunders are well known, but their lesson has not yet been learnt:

1. William Nernst wrote in his *Theoretical Chemistry* (London, 1923): "Dalton's atomic theory rose by one effort of modern science like a Phoenix from the ashes of old Greek philosophy."

2. Max von Laue, in his *History of Physics* (London, 1950), said: "The voluminous literature of atomistics extends through all the centuries, but so far as antedates 1800, and despite the famous names that occur in it, no favourable verdict can be rendered as regards its usefulness. The only exception is the quickly forgotten statement (1738) made by David Bernoulli concerning a kinetic theory of gases."

The Chronological Table in Chapter IV includes some twenty names between 1600 and 1800, three of whom made fundamental contributions to the mathematics of atomism: Huygens, Newton, and Boscovich; while Boyle and Hooke did much to strengthen the idea. Neither Nernst nor von Laue can have read Newton's *Principia* with understanding of its historical importance, for Dalton refers repeatedly to Newton's derivation of Boyle's Law. The truth is that over a hundred and fifty years of atomic speculation, observation, and calculation in physics and chemistry had prepared the way for Dalton and for XIXth-century physical and chemical atomism.

There is no doubt of the need for an up-to-date, balanced, and comprehensive work on the history of atomism, drawing ideas, mathematics, and experiment together into a single story. When available, it should become required reading for all students of the exact sciences.

For that I am not competent. This is not a history of atomism, tracing its development through the centuries. All that

I have attempted here is to present for busy teachers, research workers, and students a brief introduction to the idea of atomism and its history, including a Chronological Table, and stressing the variety of the basic conceptions and their gradual transformation. Emphasis is placed on fundamental theoretical aspects of the conception. Because the treatment throughout is in terms of basic *ideas,* I believe that a considerable part will be intelligible to readers without training in mathematical physics.

I have presented this in the way which is natural to me: as if the reader wished to view the history of atomism, not as a fully understood chapter of the past, but as part of a remarkable unfinished story, still very puzzling. If one young mind that happens to light on this essay is encouraged to meditate on fundamental problems, that is enough.

To the mature reader I say this: no one is so brilliant that he can afford to neglect what history can teach him.

CHAPTER 2

Atomism

In the broadest sense, atomism means the reduction of complex phenomena to fixed unit factors. This includes *epistemological* atomism, or the doctrine of units of perception; *linguistic* atomism, the use of an alphabet; *logical* atomism, the postulation of unit propositions; *biological* atomism, the assumption of discrete cells, genes, etc.; and various kinds of *social, economic,* and *psychological,* as well as *physical* atomism, which is our concern here. Atomic methods can be applied even where the units are not isolated entities randomly arranged, but components in an ordered system, as are the genes in the genetic system. Some kind of atomism, or use of constant irreducible units, is probably indispensable in the systematic ordering of complex facts of any kind. Unit factor statistical analysis is an atomic method applicable in many fields. Even the use of co-ordinates and differential equations—often regarded as the characteristic expression of continuity—depends on the availability of atomic systems providing invariant quantitative units or measures of length, time, etc. Moreover, atomism is often latent where it is not explicit, as in the application of group theory to symmetrical arrangements. For symmetry can be studied without explicitly considering the nature of the units whose regular arrangement forms the symmetrical pattern.

This analytical procedure, the reduction of complexity to units, comes naturally to the Western mind, and appears to be the only policy for intellectual exploration discovered so far which can be pursued systematically, as a *method*. Atomism is the pre-eminent intellectual method. If a better one lies ahead, it must grow out of, and must subsume, the achievements of atomism in physics and elsewhere.

The structure of mental processes is as yet little understood, but the rational conscious intellect, as developed in the West, is basically analytical, in the sense that it moves by definite operations or steps from one more or less isolable stepping stone (image, idea, or word) to the next. It seeks discreteness and permanence in these stepping stones, and infers such properties whenever possible. The unconscious mind may not proceed by indivisible steps, but reason must when it seeks clear discriminations. And where appropriate it simplifies matters for the intellect if the stepping stones can be treated as identical with one another—that is, as intrinsically equivalent units.

Thus discontinuity of its linguistic and logical *terms* is for the conscious analytical intellect psychologically and logically prior to notions of continuity. The intellect pays attention to and thinks about the stones *before* it becomes aware of the steps between them. This functional priority of substantive units may not have been reflected in the history of the development of reason in all human communities. For example, early Chinese thought may have reflected continuity of process more directly. But it is relevant for the West that the Pythagoreans, with their discrete integers and point patterns, came before Euclid, with his continuous metrical geometry, and that physical atomism as a speculative philosophy preceded by some two thousand years the conception of a continuous physical medium with properties of its own. Moreover, the counting of time was a commonplace before the sophisticated idea of a continuous temporal quantitative magnitude was developed. During the centuries of the Middle Ages in Europe, when the Greek atoms were almost forgotten, the Latin word *atomus* stood for the smallest unit of time: "the twinkling of an eye." Quanta of process, and their numbers, come naturally to minds which themselves move by steps, but first come the stepping stones. Thus the two atomicities of the XXth century—the discreteness of the material constituents of physical systems, the discontinuity of process and interactions—directly reflect the dual discreteness of the operation of the intellect: the stepping stones and the steps; but they may be none the less objective for that.

This being so, it is not surprising that the history of physical atomism has been one of the most important and exciting adventures of the human mind. Whatever the limitations of atomism, and however profoundly it may be modified in the future, it is here that the ordering intellect has come most closely to grips with the objective structure which, I believe, existed before there were men and will survive them. Hence the unique prestige of atomic physics; hence nuclear power; and hence also the greatest dangers confronting mankind.

But physical atomism is more than logical analysis. It is the assumption that there is a quantitative limit to division, that small ultimate units exist, and that large-scale phenomena are to be accounted for in terms of the small and therefore in terms of these ultimate units. This is the essence of atomism: the view that physical structure is not infinitely complex and that there exists a limit to research into smaller regions. Atomism has rightly been described as a policy for research; it can never be proved that the ultimate units have actually been reached, though at some stage it may become unnecessary to assume the existence of any finer structure or further complexity.

The atomic assumption has much in its favour. A survey of the *prima facie* evidence for the discrete construction of matter would cover a great part of physics, from the everyday properties of expansion and contraction, dispersion, and so on, to the subtleties of optical, stereochemical, and crystalline properties. Perhaps the most striking direct evidences of discrete structure are the Brownian motion, the close agreement of the values of Avogadro's number obtained by different methods, radioactive properties, the X-ray diffraction patterns produced by crystals, and the photographs of particle tracks. Though these phenomena do not prove that *ultimate* structure has been reached, they point unmistakably to discreteness of structure in nuclei, atoms, and molecules.

But atomism is not merely a policy or method which has proved brilliantly effective at certain levels of material analysis; it is also a positive assumption regarding ultimate structure. This assumption has a powerful psychological appeal, for it suggests a limited task with high rewards. If there really exist ultimate units, we have only to discover their laws and all their possible combinations, and we shall be all-knowing and all-powerful, like gods. So it seems.

This dream of final atomic knowledge seized many Western scientific minds, consciously or unconsciously guiding them to unexpected discoveries which in a sense justified the dream. But to others, mainly those of religious and aesthetic temperaments, the analytical assault on nature was a perversion and a sacrilege, an arrogant attempt to steal secrets for which they intuitively felt man was not worthy. For them the atomist was necessarily an atheist. Sinful man was not fit for atomic power. If it seems strange to us today that many Christian thinkers through the centuries had already reached that conclusion, this is only because the vague dream of the power of atomic knowledge has today become such a hard fact. Moreover, it was not only the Christians who distrusted atomism; there were others who felt that it was wrong to tear beautiful things to pieces, instead of accepting, enjoying, and using them as they were. Indeed many today consider that the issue is still open, and that it may be wiser to enjoy life in the south in the shade of a banana tree than to fear it, at least for one's children, in the north in the shadow of the atom.

It is hard for us today, with our nuclear preoccupations, to realize what passions the doctrine of atomism evoked long before the atom was discovered by exact scientists. Atomism originally stood for iconoclasm, impiety, and atheism, because the Greek atomists conceived a universe under the reign of chance. Indeed most atomic thinkers through many centuries neglected the order, harmony, and beauty, the adaptation of structure to function, the *ordered* dance of the atoms, which mark many aspects of the world and for the religious are evidence of a divine design. If God is a principle of order, then atomic chaos, the disorder and random collisions of material particles, are the province of the Devil, and the atomic doctrine the most dangerous of all challenges to righteousness and wisdom.

Democritan physics and neo-Darwinian biology became scandalous to most Christians as they realized that these doctrines ascribed to chance motions and mutations a primary role in the governance of the physical universe and of the history of life. Religion and science are both expressions of the desire of the human mind for harmony and order. It was therefore natural for religious minds to regard the scientific enthroning of random disorder as the supreme iconoclasm, for it betrayed the common origin of science and religion. Agnosticism could do no serious harm to religion as long as it continued to believe in order, but agnosticism based on atomic disorder was not merely an anti-Christian rebellion; it was an organized assault on all the gods, on the very idea of God, on Order itself.

Atomism certainly disrupts systems and studies isolated parts, and it must do so before it can direct its attention to the ordering relations of the parts in the various natural systems, and learn to put them together again as they were. It is, I suggest, for the atomists to silence their critics by achieving what is a long and hard task: the identification of the atomic orderings in everything, from nucleus to cosmos, though in doing so atomism will acquire a new significance. Indeed this task has already been begun: the *arrangement* of the particles in nuclei, atoms, molecules, crystals, viruses, organisms, and universe, is now the centre of attention. The science of "atomism" is thus becoming the study of structure, and structure is the relations of parts viewed as a form of ordering. The consummation of the atomic tradition may be to issue in a *general theory of order and disorder,* of structural transformations and stability, placing less emphasis on the units and more on the observed structure of relations. This is not an empty speculation, for it is already happening. The elementary particles of 1930–1960, or most of them, depend on their setting; they are tiny functions of much larger environments, their classical material properties have almost vanished, and when grouped they possess collective properties.

Our elementary particles are more elusive than Democritan atoms, but in some sense "they are there." What counts for physics, and for us in this essay, is not the degree of their "objective reality" but their scientific role as elements in a representation of observed relations. As Ernst Cassirer put it, what matters is "not the existence of things, but the objective validity of relations." But the *nucleons* are as real as anything is, for of all the repertoire of fundamental particles they provide a relatively stable "material" basis for the indispensable apparatus of physics: the rods, clocks, and focussing instruments of all kinds. All the other particles are anchored to nucleons; it is the nucleons which for practical purposes fix the initial conditions of all experiments.

Two extreme interpretations of atomism have persisted through centuries: the naïve assumption of objectively real indivisible pieces of matter, and the sophisticated view that "atom" is merely a name given to abstractions which it is convenient to assume in simplifying complex phenomena. The first certainly goes back to Leucippus and Democritus. The second perhaps stems from Ockham, who wrote in 1330 of "the fiction of abstract nouns"; from John Troland, who in 1704 interpreted material particles as mental fictions; and from countless others down to Ernst Mach, who after starting as a physical atomist came to regard atoms as "mental artifices" or "economical ways of symbolizing experience."

Both views have advantages: the first provides a familiar image that holds the attention, the second suggests that such images are not as simple as they seem but require analysis. For an idea like that of a "hard particle" may evoke a satisfactory visual picture and yet be logically and mathematically vague and obscure.

Why, then, has this particular obscure idea, the notion of "atoms," enjoyed such a vogue and so great a success? To answer this question we must consider what physics is.

"Science is finding things out" (Collingwood), and I shall assume that in quantitative physics this finding out takes the form of the cumulative, self-correcting, systematically pursued interplay of ideas and measurements, expressed in a mathematical theory. Physics is thus the continuous critique of quantitative laws. Theory confronts experiment, and both sides are a mixture of obscurity and clarity. On the one hand, theory uses familiar obscure ideas which suggest a precise quantitative calculus; while on the other, experiment uses obscure objects (material apparatus, sources of energy, etc.) whose laws are not yet fully understood, to obtain precise quantitative measurements. Unconscious assumptions are hidden in the obscure ideas which have suggested the calculus, and unknown structures and processes are exploited in the experimental situation which provides the numerical results. For example: matter, space, time, atomic particles, etc., are familiar obscure ideas, which have suggested clear calculi, such as coordinate geometry, matrix theory, etc.; while every experimenter, in setting up his apparatus to obtain accurate observations, uses systems whose laws are still incompletely known. Thus the interplay of ideas and measurements improves and extends both, by drawing previously concealed features out of the obscure realms into the clarity of mathematics and of measurement.

This systematically exploited two-way interplay is characteristic of science, though its dual character has sometimes been neglected. One school of scientists and philosophers has overstressed the *a priori* quasi-deductive aspect of physical theory. Another has placed an equally misleading emphasis on the observations from which generalizations are supposed to be derived by quasi-inductive procedures. Both these schools have been at fault, for they have neglected the complex process of observation and cognition, the interplay of idea and observation, each serving to guide the other to new achievements by a formative mental process which is neither deduction nor induction.

The more balanced interpretation is old. Diderot, the French encyclopaedist, expressed it in 1750: "But are experiments made haphazardly? Is not experimentation often preceded by a supposition, an analogy, a systematic idea that experiment will either confirm or destroy?"

And Mach wrote in 1903: "There is no pure experimental research for, as Gauss says, we actually always experiment with our thoughts."

The history of the conception of atomism illustrates this interplay more forcibly than that of any other scientific idea. At no stage was the idea of an atom, or ultimate particle, either an arbitrary human invention or a logically necessary inference from observation. It was always an intelligent invention and a plausible inference. Moreover, the conception of minute ultimate units not only set an aim for research promising unprecedented rewards, but demanded increasingly accurate techniques. No challenge to ingenuity which man has yet created for himself surpasses that of acquiring the absolute knowledge implied in the existence of fundamental structure. The Greek atomists dimly conceived this possibility, but they did not appreciate that steadily increasing accuracy of measurement, quantitative precision, would convert the plausible inferences of the Vth century B.C. into the all-disintegrating pneumatic drill of the XXth A.D.

It was natural that Greek atomism should start from a non-quantitative analogy: Since most bodies are divisible, are not all bodies? This provoked the more subtle and fertile question: Since visual perception has a limit, may not physical divisibility also? That second question was provocative; it held the attention of the mind, giving it a new basis for speculation: the conception of a rock-bottom minute world which constituted the real basis of everything.

The Greek schools (often wisely relaxing in their gardens —they lived long lives without medicine) debated the philosophic question: Do ultimate atoms exist? Many (but not all) physicists and chemists of the XIXth century claimed to give the answer: "Yes, chemical atoms are indivisible and eternal." In the same period a more cautious group preferred not to use the term "atom"—from ἄτομος, the indivisible—for Dalton's chemical units. They were right, but the name stuck. And now, in the 1960's, the question has become ambiguous. What does it mean to "divide" a quantized-field-particle which extends to infinity but may vanish at any moment? Are *baryons,* which in a technical sense are localised and conserved, the XXth-century "constituents" of material systems? Or nucleons? We cannot say for sure. The old-style spearhead is no longer sharp, but we do not know what new design is needed.

This provokes the awkward question: If atomism is always changing, what characteristic must a physical theory possess if it is properly to be called "atomic"? What is this essay about?

The history of science shows that sharp definitions lead to trouble, often quickly. Dogmatism in science is usually mistaken, because the conviction of certainty expresses a psychological compulsion, never any truly compelling reasons or facts. When a view attains wide popularity and seems obviously beyond question, its decline has usually begun or will begin very soon.

In 1892 W. W. Rouse Ball of Trinity College, Cambridge, who was well informed on competent opinion—J. J. Thomson was also at Trinity—wrote: "The popular view is that every atom of any particular kind is a minute indivisible article possessing definite qualities, everlasting in its properties, and infinitely hard." Rouse Ball wisely added descriptions of two rival atomic doctrines: Boscovich's point centres, and another based on twists in an elastic solid aether.

Four years later the hard everlasting atom began its rapid exit from physics. In 1896 Becquerel discovered the radioactivity of uranium; in 1898 the Curies found the same property in thorium and radium; and in 1902–1903 Rutherford and Soddy proposed the transformation hypothesis. The "popular view" barely survived a decade.

One might define as "atomic" *any theory based on the changing spatial relations of a finite number of separate indivisible entities of as few kinds as possible.* But what is an entity? How permanent must it be? Are quantized fields "separate indivisible entities"?

Another tempting definition is: *Any theory is atomic which assumes that a finite number of ultimate entities are present in any finite system.* This stresses the spatially finite character of the relations of the particles, but it solves no problems and leaves the status of quantum mechanics obscure. Fertile theories often elude supposedly clear definitions.

It is more profitable not to attempt a general definition and instead to consider what conceptions of atomic particles have been most fertile and have had the widest applications. There have been only three basically distinct and widely successful conceptions. They present striking contrasts.

A. *Democritan-Newtonian hard atoms.* Newton (c. 1686–1720), following the Greek atomists, adopted as most appropriate the view that the ultimate constituents of matter are themselves minute, hard, permanent, indivisible bits of matter of definite sizes and shapes. They occupy parts of space and move through the empty spaces between them. Occupied and empty space are sharply distinguished. Many scientists still think like this.

B. *Boscovichian point centres.* Boscovich (1758) proposed a substitute having several advantages. The smallest units are not extended pieces of matter, but persisting physical points evidenced as centres of interaction. Only the spatial relations of these physical points enter physics, and every pair of such points are in interaction.

C. *De Broglie–Schrödinger wave-particles.* De Broglie (1923–1924) and Schrödinger (1926) introduced a third, more powerful, conception: that in certain circumstances the ultimate particles possess wave as well as particle properties. This idea, though substantially modified in statistical quantum mechanics, is the most comprehensive of the three and provides the basis of current atomic physics. The wave theory of light had long presented a challenging contrast to atomism; here the waves seem to have beaten the particles.

"A" is familiar but mathematically awkward, involving discontinuity at the surface of the atoms. It is also inescapably associated with the idea that "matter cannot act where it is not." Even Newton believed that all genuine physical action was by material contact, though he saw no way of using this to account for his law of gravitation.

"B" is ideal for many purposes, and mathematically convenient. It implies that each point atom pervades space, acting everywhere except where it is itself. A valuable emphasis is thus placed on interactions, or changes of relative velocity.

But this idea has proved too restricted, at least as Boscovich presented it.

"C" is the most successful so far, but it lacks both visual immediacy and mathematical simplicity. Moreover, recent mathematical modifications have made its physical basis obscure. This has led to many different interpretations being offered, of varying degrees of "orthodoxy."

To these three primary ideas must be added various special conceptions of less value: vortex rings and other rotating units, twists in a rotationally elastic jelly, dislocations or holes in a close packing of spheres, negative atoms of various kinds, and so on. There is also Eddington's ghost-particle, from which he excluded all vestiges of materiality, so that it became merely a "carrier of variants" or "a conceptional unity whose probability function" is specified by certain wave vectors. Finally, recent work has suggested that what marks a "fundamental" particle is not its physical indivisibility, but the possession of a definite set of *fixed parameters* (mass, charge, spin, etc.) and no internal *variables* that might represent changes of inner structure. For a fundamental particle must be incapable of internal modification.

"C" is the most recent and certainly the most successful, but in many respects it represents a departure from the classical atomic tradition. Moreover, the high-energy wave-particles discovered since 1930 are intuitively felt not to be true "constituents" of material systems. Underlying this judgment lies a criterion which has seldom been made explicit: A true constituent of physical systems must possess characteristic properties which are *not* functions of position (relatively to other particles) or of time (relatively to events in its neighbourhood). This criterion of genuine constituents corresponds to

Epicurus' "impassibility" of the atoms, the property of not being affected by anything. Any particle whose properties, for example its times of appearance and disappearance, are functions of its space-time position relatively to the experimental system may be called *virtual* in the sense that it behaves in certain respects like a constituent particle, though not actually one. It seems that many of the recently discovered particles are virtual in this sense; they possess variable properties which are functions of their environment.

Particles which are unstable are not less "real" or "important" than true particle constituents, and they are certainly more interesting, just because they are less understood. But they lack one important feature of the constituents of an ideal classical particle theory: the determination of a unique continuity through time of localization in space, which was provided in classical theories by the persisting identity of a particular particle following a precisely determined spatial path.

The necessity in classical physics for this type of temporal continuity of spatial localization was made explicit by Hertz, who developed (1890) a system of classical mechanics in which he defined a *material particle* (for the purpose of a *macroscopic* theory) as "a characteristic by which we associate without ambiguity a given point in space at a given time with a given point in space at any other time." Boscovich's permanent point-particles were an attempt to provide the same property in a *micro* or atomic theory.

Quantum mechanics and quantum experiments have eliminated this stabilizing feature by undermining exact determinism, localisation, particle identity, and permanence. Only a certain probability remains of a one-to-one association of any spatial feature *now* with a similar feature *a moment later*.

It is sheer luck, in a sense, that any physical apparatus stays put, for the laws of quantum mechanics allow it a finite, though small, probability of dispersing while one is not looking, or even while one is. A continuous particle-track and a stable apparatus possess a finite probability and no more. For the high-energy quantum physicist, the classical particle has vanished irretrievably—or so it seems.

However, this disruption of the classical particle is not solely the responsibility of recent quantum theory and experiment, for a difficulty had arisen earlier in relativity theory. Einstein's aim was a comprehensive relativistic theory, and he knew that classical particles were alien bodies which should be represented, if possible, as unique regions in a relativistic energy field. But it was unlikely that such virtual particles or field foci could be permanent. Indeed it seems that at any time after, say, 1910 or 1915 (when it became clear that negatively and positively charged particles had definite radii associated with them) Einstein might have predicted the existence of unstable particles by using the following syllogism:

1. The laws express, in terms of intervals, the properties of relativistic space-time fields.

2. Experiments have shown that some electrical particles have an effective finite radius.

3. Therefore some electrical particles must have an effectively finite life-time.

Or, in dimensional terms: Since l_o enters electrical particle theory, $t_o = l_o/c$ must enter relativistic particle theory as a life-period (since the interpretation of t_o as the reciprocal of a frequency was not possible in this context).

Behind this line of reasoning lies the assumption that particles are quasi-localized semipermanent forms of relativistic field energy, their radii, lives, and masses being properties of temporary energy distributions which for undiscovered reasons are centred at points and which appear therefore as "unstable particles." There is much evidence for this basically anti-atomistic view, in which there are no truly discrete entities but only a continuum of transformations. The power of the energy concept, of relativistic methods, and the discovery of many unstable particles, all speak in its favour.

Yet there is a snag. However much the field physicist tries to inhibit unique points, they haunt him like a naughty remnant of particle manners. Can physics do without the sharp localization in space of centres of action which persist and form stable patterns? If everything is an energy flux, why should any definite forms exist? If the universe is a mingling of probability clouds spread through a cosmic eternity of space-time, how is there as much order, persistence, and coherent transformation as there is? Physics has little use for a pointless world.

The point-centre conception of atomic particles is suitable for a particle theory, but in a field theory it gives rise to mathematical troubles, for the field becomes infinite at the point-centre. But field theory has not yet found any adequate substitute for these unique points.

Moreover, in the background pure mathematics whispers a warning: It is easier to develop a continuum from points than to start with a continuum and to discriminate points within it.

Such general reasoning might carry little weight if it stood alone. But the empirical evidence seems to support it.

Even if we neglect the discrete mass spectrum and lives of the known particles as too much of a dark horse to be included, the remaining fundamental constants tempt us to bet 2 to 1 on ultimate discreteness of quantities:

For continuity	*For discreteness*
c, velocity of light	e, electron charge h, Planck's constant

Moreover, a closer examination removes the apparent conflict in the evidence: c is directly observed as a macroscopic ratio of space and time measurements; e and h are inferences regarding micro-phenomena drawn from macro-observations. Thus we have one observed macro-constant voting for continuity, and two micro-inferences for discrete units. There is a hint here that continuity may be in some sense more a macro-property, and that the primary micro-properties may be more discrete, than is suggested by the success of relativistic theories. In fact the case for an irreducible element of ultimate micro-discreteness remains strong, in spite of the wide validity of relativistic expressions.

Classical atomism has certainly encountered insuperable difficulties, and radical modifications have already been necessary. This is not surprising, for unlike the macro-sciences atomism is always in the awkward position of inferring the unknown minute from the known large. What is primary in direct observations can never be all that is primary in atomic theory. But in exploring the core of the atom, and very high energy processes, atomic physics is not, as has been recently suggested, exploring an alien world with *wholly* inappropriate ideas. As we have already seen, if a unified theory is possible the laws which cover the new micro-world as well as the old macro-world must be generalizations of the known laws, to be reached by dropping unnecessary restrictions concealed somewhere in the quantum mechanics of 1929. The theoretician is not *completely* in the dark in exploring the nucleus, but his lamp is of the wrong colour to illuminate its structure, and he knows this or should know it. What he has to do is to modify his lamp to give a wider band of colours to catch the unfamiliar paintings in the caverns of the nucleus.

Thus the challenge of the unknown in high-energy particle physics can, in principle, be met in two ways: by new experiments guided by speculative ideas, and by historical research to discover in the past just why and when and how physicists assumed that a lamp of a particular colour would do, and neglected other possibilities. Was it Newton, or Boscovich, or Schrödinger who took "the wrong turning"—i.e., a narrower path when a broader or more general one was possible and would, say in 1960–1970, prove necessary? The observed universe is certainly a flux, but it also possesses structures which persist and recur. What kind of atomism is necessary to describe these structures, and did anyone in the history of physics conceive the right kind of atomism but turn it down for reasons which can now be seen to be inadequate?

Most philosophers and scientists have agreed that, to account for what is observed, it is necessary to infer entities which are not observed. An unseen universe is necessary to explain the seen. The flux is seen, but to account for its structure we infer particles of various kinds to serve as the vertices of the changing patterns of light and shade, of colours bright and dark, of radiant energy. Some readers of this essay will probably live to see a new picture of this unseen world. They are to be envied.

These are the broad issues which throw light on the historical sequence of names, dates, and special achievements given in the next two chapters. But it must be remembered that while the Greek conception of an atom can be described in a few sentences, the XIXth-century atoms, and still more the mid-XXth-century elementary particles, acquire their character not from a few words or from a visual image, but from complex mathematical theories associated with physical principles. The naïve visualisable Greek atom has gradually been transformed into an intellectual focus where experiments, images, mathematics, and principles converge, and this complexity cannot be covered in a brief survey. We shall be compelled largely to neglect those broader issues which, in each decade after, say, 1850, give the particles their opportunity and their characteristic form:

Space, metrical geometry, coordinates.
Time, reversibility, irreversibility.
Inertia, mass, force.
Energy, entropy.
Statistics, probability.
Symmetry, invariance, asymmetry, chirality.*
Simplicity, complexity.

* "Chiral" (from χειρ, hand) means "non-superposable on its mirror image," like right and left hands.

CHAPTER 3

History and Geography

THE history of atomism is highly complex, and the simplifications offered here must be treated with caution; for:

Logic tidies up retrospectively the work of individuals whose assumptions were obscure and whose ideas were often confused.

Individuals are remembered for their most influential ideas. Opposite views may have been held at different times by the same person, and may be simultaneously dominant in different sections of one community. Even the critics of atomism, such as Aristotle, Galen, Goethe, Ostwald, and Stallo, called attention to the idea and by their comments often contributed to its development.

Discoveries may be accepted only after a lag. Maxwell's theory of the electromagnetic field was not regularly taught in Continental Europe until thirty years after its publication.

We shall begin with broad perspectives. Physical atomism can be said to have started with Leucippus and Democritus, and after many transformations to have reached a "crisis" (as it seems to us now) after 1930.

This span of 2400 years may be viewed as the life-history of an idea which is born, matures, achieves its task, perhaps then marries another idea, and leaves its progeny to carry on. The analogy has some value. Certainly in the case of atomism there were moments of conception, and periods of quiet growth, of "play" (spontaneous qualitative speculation), of "work" (deliberate application to quantitative tasks), and of fusion with other ideas. There were phases of high activity and others of passivity or apparent decline. Table I illustrates this analogy.

TABLE I

Life-History of the Idea of Atomism

B.C.	450– 420		The idea is conceived,
	420– 280		and welcomed.
B.C.	280–1400	A.D.	It sleeps, with occasional play.
A.D.	1400–1600		More vigorous play.
	1600–1800		It is put to work quantitatively.
	1800–1900		The chemical atom is established,
	1900–1930		but proves misnamed.
	1930–1960		Unstable progeny confuse the issue.

Until 1900 A.D. this story took place on a small stretch of land running northwest from India through Greece and Western Europe to Britain, for atomism never took root in China or Japan. If Indian speculations are neglected, this area contracts to a tiny strip, a mere fiftieth of the earth's surface, extending from Abdera in Thrace to Edinburgh. This was the soil, and the seed drifted along it northwestwards, from Greece and the Mediterranean through Italy and Spain to northwest Europe: Britain, France, Holland, Switzerland, and Sweden. The German-speaking peoples made a brief but important contribution to basic atomic theory between 1850 and 1933.

After 1900–1910 the seeds spread westwards to North America, producing in the U.S.A. an unprecedented empirical harvest. But no fundamental theoretical advance of the first magnitude, worthy to rank with those of Planck, Einstein, Bohr, Schrödinger, or Heisenberg, has yet been made by anyone born outside Europe, apart from the New Zealander Rutherford, perhaps the last earthy Democritan to fertilize physical theory.

TABLE II

Geography of Atomism

The records do not permit any conclusions regarding atomic ideas before 600 B.C., e.g., in Egypt or Sumer.

B.C. 600

400 GREECE. Speculative, non-quantitative atomism on rational casual basis.

200

ITALY. Romans honoured the Greek ideas.

A.D. 200 INDIA. Early speculative ideas, mainly intuitive and linked with religious attitudes.

400

600 MEDITERRANEAN. After the decline of Rome, Arab and Jewish thinkers maintained the memory of the

800 Greek ideas and developed some variants.
EUROPE. Some Christian scholars showed interest.

1000

1200

1400 ITALY. Revival of speculative interest. Italy and Spain served as channels for the European rediscovery of Greek science.

1600 WESTERN EUROPE. First quantitative applications made in Italy, Britain, France, Holland.

1900 WESTERN HEMISPHERE. Experimental activity spread westwards, though basic theory remained mainly European.

Women and Orientals have so far made relatively minor contributions, but this will surely change. For since Marie Curie in 1898, Irene Curie, C. S. Wu, and other women, as well as Yukawa, Tomonaga, Nishijuna, Namba, Oneda, Lee, and Yang, have blazed the path.

Table II summarizes the geographical aspect of the history of atomism.

In this essay we are primarily concerned with the *basic ideas* of atomism. Though a great number of individuals have made experimental or theoretical contributions, and of these over a hundred are mentioned in the Chronological Table, only five or six appear to me to have made genuinely fundamental theoretical advances in improving or clarifying the general physical conception of ultimate atomic particles since the foundation of modern quantitative physics by Galileo and Kepler. I consider that the following contributed crucial ideas which cannot be replaced by the work of others:

NEWTON c. 1686. Strengthened and rendered partially quantitative the Greek conception by introducing attractive and repulsive *forces,* stressing the shapes and properties of ultimate particles and estimating their size.

BOSCOVICH c. 1758. Blended atomic and quantitative methods in a mathematical scheme for a unified theory of physical *point-centres.*

DALTON Dalton (1803–1808) identified the *chemical atom,* showing how to assign quantitative prop-

PROUT erties to chemical atoms; and Prout (1815) suggested that all atoms were composed of *hydrogen atoms.*

DE BROGLIE De Broglie (1923) and Schrödinger (1926)

SCHRÖDINGER developed the conception of *wave-particles.*

These key contributions to the idea of atomism provided the basis for a large number of special applications, which can roughly be grouped as follows, though their interactions were profound:

From 1800: Chemical properties.
     1820: Kinetic theory of fluids, and later of solids.
     1830: Electrochemistry.
     1869: Periodic table.
     1880: 230 crystal groups, optical properties.
     1890: Electron theory.
     1900: Relativity theory and quantum theory of particles, their wave-fields, combinations, and interactions.

These are the broad vistas. But "nothing comes from nothing." How and why did Greek thinkers come to initiate this adventure of thought, and later of experiment? What intellectual or psychological factors guided their thinking?

Classical scholars—for example, Bailey in his *Greek Atomists and Epicurus* (1928)—have interpreted Democritan atomism as a resolution of philosophical dualisms which had occupied earlier Greek thinkers: the relations of the One to the Many, of Unity and Variety, of Permanence and Change, of Reality and Appearance. Leucippus and Democritus achieved this resolution through a new and sharp dualism constructed for the purpose: *Atoms* and *Void*. Democritus knew that he couldn't build a cosmos from sense data alone, so he accepted the atoms and void of Leucippus. At the level of conscious rational analysis this interpretation need not be questioned, but it leaves the XXth-century scientist unsatisfied. Beneath these multiple philosophical concepts, were not more general and perhaps scarcely conscious factors at work which led the speculative reasoning of Leucippus and Democritus to an idea of such extraordinary power? What were the factors which gave atomism its unique status for the scientific intellect in the realm of quantity? What made quantitative atomism at once so comprehensive and so objective?

The germ of an answer has already been suggested in general terms: Atomism is a direct expression of basic tendencies in the Western analytical intellect. But this does not explain why a dualism, and why just *material atoms* separated by *void*. Here the historical facts can help.

In the XVIIth century it was widely believed (see, for example, Gale's *Court of the Gentiles,* 1667–1671) that atomic ideas were of high antiquity, traceable back from Democritus to Pythagoras, to Moschus (a Phoenician), and to Greek, Chaldean, and even Hebrew thinkers mentioned in the Old Testament. We need not accept that particular myth. But there are definite signs, for example in Sumer and Egypt many centuries before Pythagoras and Democritus, of germs of physical atomism, of the conviction that the universe is constituted of discrete parts which can be observed and sometimes even counted: *grains* of dust, *drops* of water, *seeds* of life. Similar ideas are found in early Greek documents, and the conception of smallest particles of living matter has had a long history.

But these were exploitations of special ideas which could not be of general application; they could not suggest, for example, why drops are not seeds and seeds not grains. For this it was necessary to renounce these exhausted special ideas and go back, as it were, to a more general idea, logically implicit and unconsciously latent in these special ideas. Here lay the genius of the Greek atomic school. It discarded drops, seeds, and grains as too special and went back, or deeper, to a hypothetical universal unit of which all others could be composed, a fundamental existent which does nothing but fill space where it is, and move to where it is not. Through this emancipation from preoccupation with special units a new vista is opened to thought. *All* the processes of the universe, and all its varied forms, can now be regarded as nothing but the arrangements and rearrangements of these universal units, the ultimate material atoms.

We need not trace in detail how this novel and expressly invented dualism of occupied space called Matter, and empty space called Void, appeared to solve the various problems posed by the earlier Greek thinkers. All the earlier philosophical dualisms were, in fact, subsumed under one new comprehensive dualism which appeared to be, and was in fact, susceptible of objective scientific exploitation. For science at that time and long afterwards it was a happy chance that Leucippus and Democritus rejected any divine ordering principle guiding the atoms towards their ordained places, and substituted a natural causal necessity evidenced in their movements.

If genius is the creation of what is timely and fertile, there is no doubt of it here, the more so because this Matter-Void dualism is not *necessary,* either for the human mind or in relation to the observed facts, as we shall see. But it was the best answer to many basic questions which the human mind was able to conceive from 450 B.C. to 1750 A.D., for nearly a hundred generations.

What is familiar and obvious to us was to the Greeks of 450 B.C. a shocking novelty in the form of a strange, unbalanced dualism: physically real finite units of material separated by also "real" but intellectually empty void. The empty space had to be there, but only passively, to allow the atoms to move. This was more than a speculation regarding the constitution of matter; it was a universal doctrine regarding the ultimate nature of the cosmos, with many contrasted implications for different types of thinkers. But beneath these religious, aesthetic, and philosophical implications, it appealed to all genuine enquirers as offering a powerful instrument in the search for objective understanding of the universe in which they found themselves.

It is interesting that the intellectual importance of atomism was so clearly recognized, even though Leucippus and Democritus not only neglected but even partially rejected the Pythagorean-Platonic doctrine that mathematics held the supreme key to understanding. Thus the Greek atomists, occupied with their philosophical rebellion against an inadequate religious tradition, preferred to emphasize the supposedly random chaotic motion of the atoms, than to accept (prematurely, as it would have proved) a quasi-religious faith in mathematical order. There is no dichotomy of black and white, or of useless and useful, in the history of science; it is always a question of where and when and how. What we crudely label as mysticism and logic lie together at the sources of the mind, and each has its time and place.

At the time of the Greeks, atomism did not need and could have made little use of Pythagorean principles. It was designed to provide a qualitative interpretation of well-known phenomena such as expansion and contraction, solution, dispersion, precipitation, and so on. Only its manifest success in these relatively simple matters enabled atomism to acquire the clarity and reliability which were prerequisites for its application to quantitative theories of other mechanical, chemical, thermal, optical, and electrical properties. But Greek atomism was potentially quantitative, for occupied and empty space invited treatment by geometry.

As we shall see, it was the combination of *two* keys to the understanding of nature—Pythagorean *Number* and Democritan *Atoms* (or *three* if we include Euclidean *Geometry*)— that led to the atomic theories of the XVIIth century and onwards. Though we are here more concerned with the physical ideas than with their application in mathematics, without the passionate belief in mathematics of the Pythagoreans and their followers through the centuries there might have been no exact science as we know it. The satisfying Pythagorean representation of numbers as plane patterns of dots, and the fascination exerted by regular three-dimensional patterns of points—such as the vertices of the five regular solids—on such minds as Plato, Leonardo, Kepler, and many others, led directly to static geometrical atomism, the interest in regular atomic arrangements in crystals and elsewhere. However, Democritan atomism was primarily concerned with change and movement, and Huygens, Newton, and Boscovich were respectively led to explore the possibilities of a kinetic, dynamical, and kinematic atomic physics.

No tidy classification of the ideas or analysis of the story is yet possible, for unresolved contradictions are still present today. We are not yet wise enough to analyse the historical record in any final manner. But it is probable that both the successes and the weaknesses of XXth-century atomic physics arise from the unresolved tensions of this unique triangle, not domestic but cosmic: Pythagorean and Euclidean *Number* invading the Democritan marriage of *Atoms* and *Void*. For Number, Particle, and Field have not yet discovered how to work together properly in the world as it is.

CHAPTER 4

Chronological Table

T HIS table does not provide a history of atomism, for no attempt is made to describe the background of each thinker, to analyze his achievement, or to show how one step led to another. I started preparing it some years ago because I was disturbed that important names were in danger of being forgotten and that working physicists might accept too narrow a view of the history of such an important idea.

The aim of the table is thus to serve as a record and reminder of the most important names and dates, with a brief hint of each achievement, emphasizing ideas rather than experiments, which are amply treated elsewhere. In order to provide background I have included a few marginal names, even one or two who made no contributions to atomism. For the history of atomism cannot be understood unless it is realized that some Europeans of the highest genius, intensely interested in nature and aspects of science, did *not* recognize the importance of atomic ideas, Leonardo and Goethe for example.

The table may be either read through or used for reference. The reader who does not wish to read it consecutively will find breaks for discussions at the following points:

1650 Quantity first applied to structure of matter
1700 Newton

1750   Boscovich
1800   Dalton
1840   Faraday
1880   The climax of classical atomic physics
1900   The origins of relativity and quantum physics
1925   The years of triumph

B.C.

600
550   Pythagoras (fl. 550) visited Egypt and Babylonia as a young man and was probably the founder of the Pythagorean mathematical school at Croton (Italy), which held that number is of the essence of things. They represented numbers as patterns of
500   points.

           Anaxagoras (498–428) considered that the universe was originally a chaos of infinitely small seeds.

450   Leucippus (fl. 450)

           Leucippus and his pupil, Democritus, created at Abdera (Thrace) the Greek atomic school of natural philosophy. Democritus, the more systematic thinker, extended his teacher's ideas. The infinite universe consists of ultimate indivisible entities, all hard, permanent, and unchangeable, of one

Democritus (fl. 420)

          homogeneous substance but of various shapes and
400   sizes. These *atoms* are in ceaseless movement under causal necessity, vibrating and whirling, though sometimes fitting together in stable combinations. All the variety of the universe results from the differences in size, shape, position, order, and motion of these ultimate atoms of a single substance. The atoms of fire are smooth spheres. Other members of the school developed these ideas in various
350   manners.

B.C.

Plato (427–347), influenced by Pythagorean and Democritan ideas, emphasized the importance of geometry, and in the *Timaeus* gave a geometrical representation of the particles of fire, earth, air, and water, in terms of four regular solids: the tetrahedron, cube, octahedron, and icosahedron.

Aristotle (384–322) recognized that the Democritans had made a rational attempt to interpret change, but his own doctrine of substances and essences was antagonistic to atomism. His suggestions regarding *natural minima* (smallest particles of various kinds) are both vaguer and more specialized than the Democritan ideas. Aristotle's references to Democritus preserved the memory of Greek atomism during the Middle Ages, and are one of the main sources.

EPICURUS (341–270)

The Democritan school at Abdera had little immediate influence, but Epicurus reformulated and
300   extended their ideas, as a method of inference from the visible to the invisible, in his influential "Epicurean" philosophy.

250
200   Around 200 B.C. interest in atomism began to decline, though the Epicurean philosophy prevented it from being forgotten.

150

Asclepiades (b. 124 B.C.), Greek physician in Rome, ascribed diseases to alterations in the size, arrangement, and motion of the atoms in the body.
100

LUCRETIUS (98–55)

Titus Lucretius Carus, Roman poet and philosopher, gave eloquent expression to the Democritan

B.C.

50   doctrine in his *De Rerum Natura,* this poem being the principal channel by which interest in Greek atomism was preserved. "Part of this poem is an orderly treatise in physics" (Andrade).

A.D.

1

Hero of Alexandria (fl. c. 25 A.D.), writing in Greek, emphasized the importance of the size and shape of the vacuum or empty space between the particles of bodies, his ideas influencing Galileo, Bacon, and Descartes.

100   "Kanada": traditional name associated with the Hindu Vaiśesika Sūtras (probably recorded during the first two centuries A.D.) which, like the earlier Jain Hindu writings, contain speculations on atomism, possibly influenced by Greek ideas.

Galen (fl. 170), Roman physician, opposed Asclepiades' views on atomism.

200   St. Dionysius of Alexandria (b. c. 200) stated the Christian objection to Greek atomism: that it lacks a principle of order. "Who, then, is it that discriminates between the atoms, gathering them or scattering them . . . ? And if the combination of the atoms, as being soul-less, was unintelligent, they needed an intelligent artist to put them together."

300

From 200 to 1400 some awareness of Greek atomic ideas was maintained by Jewish and Arab atomic speculations, by Christian criticism, and by grammarians in the monasteries. Copies of Lu-

400   cretius' poem were preserved in monastic libraries through the centuries; two excellent IXth-century manuscripts exist today.

500
600   Isidore of Seville (560–636), encyclopaedist, reported on atomic ideas, distinguishing "atoms" of matter, of time, of number, etc.

700
800
900   Rhazes (865–924), Arab physician and alchemist, adopted a type of atomism. Around this time several other Islamic thinkers developed atomic ideas similar to the earlier Indian speculations. These included point atoms which kept a finite atom of space occupied, and a doctrine of "time-less leaps," discontinuous both in space and time. One school of Arab philosophers maintained an atomic doctrine for several centuries (c. 900–1200).

1000
1100

           Maimonides (1135–1204), Jewish philosopher born in Spain, reported the Arab views on
1200   atomism, exerting some influence on thought in Europe.
1300   Nicolaus of Autrecourt (c. 1300–1350), a critic of Aristotle, considered the motions of atoms under their mutual attractions.
1400

           A search for classical manuscripts made by Poggio (1417) in monastic libraries led to the discovery of a copy of Lucretius' *De Rerum Natura,* which was sent from Germany to Italy, printed in 1473, and soon became well known.

Nicholas of Cusa (1401–1464)
           Nicholas of Cusa drew attention to Greek atomism and supported it with new arguments.
1500   Leonardo da Vinci (1452–1519) in a vague sense saw mathematics in all phenomena and was

fascinated by the five regular solids, but was not interested in atomism.

Copernicus (1473–1543) wrote: "The minimal and indivisible corpuscles, which are called atoms, are not perceptible to sense . . . but can be taken in such a large quantity that there will . . . be enough to form a visible magnitude."

Thomas Langley (1546) referred to the Democritan atoms and void.

1550   From 1550 onwards, steadily increasing interest was shown in Greek atomism and its relation to the new sciences of mechanics, chemistry, etc. In 1575 F. Commandinus published his influential translation of Hero's writings. By around 1650 it was customary for learned writers to assume that their readers knew that Democritan atomism was one of the most important theories of the physical universe.

Giordano Bruno (1548–1600)

For Bruno, who was well read in Greek thought and who discussed Lucretius, the physical world consisted of *discrete minima* in eternal motion, but these were rather a requirement of thought than a matter for investigation.

Francis Bacon (1561–1626)

Bacon supported the ideas of Democritus and Hero where use could be made of them, suggested some modifications, and regarded heat as a motion of the ultimate particles of bodies. "The theory of Democritus relating to atoms is, if not true, at least applicable with excellent effect to the exposition of nature." Atoms are either "the smallest portions of matter" or "solid particles, without vacuity." ". . . if we could discover the original or primary component Particles of Matter, so as clearly to discern their Arrangements and Compositions, whereon the

Form or properties of different Bodies depend . . ."
Galileo Galilei (1564–1642)

1600

Galileo read Democritus and Hero and broadly accepted the Democritan ideas, applying them speculatively to heat and light, suggesting that the motions of the particles affect the physical properties of bodies, and separating spatial relations and motions from perceptions of colour, warmth, etc.

Kepler's outlook was Pythagorean; he considered (1611) that the hexagonal crystals of snowflakes might be composed of densely packed spherical globules, though he did not relate this to Greek atomism. He believed in a psychic atom, or monad, influenced by the planets.

In 1599 Shakespeare used the phrase "to count atomies" as a symbol for something beyond human power.

In 1601 Nicolas Hill, an English philosopher, published a discussion of the Democritan and Epicurean doctrines.

1610

1620

One of the earliest expressions of the idea of composite systems of various orders—i.e., of first order, second order, third order, etc., combinations of atoms—appeared in a work by the French physician S. Basso (1621).

In 1624 the Paris Parliament decreed that persons maintaining or teaching atomism, or any doctrine contrary to Aristotle, would be liable to the death penalty.

1630

From around 1630 "atomism," in the more general sense of material particles not necessarily indivisible, became the dominant theory of matter.
D. Sennert (1572–1644)

Sennert, German physician, published (1636) the first systematic application of atomic ideas to chemistry (e.g., that atoms can be recoverable from their compounds, gold from solution in an acid), being followed by other chemists (J. Junge, E. de Claves) anticipating Boyle.

M. Mersenne (1588–1648) accepted a Democritan type of atomism, but as a working hypothesis, not a materialistic philosophy.

1640 Descartes (1596–1650) discarded the Democritan Matter and Void, and replaced them by a medium (the aether) with its own properties, composed of infinitely small particles in motion; matter was for him *spatial extension*. Though Descartes did not believe in indivisible particles, his mechanical doctrine exerted great influence.

In 1646 J. C. Magnenus, a Frenchman living in Italy, published a detailed account of Democritan atomism, the reasons in its favour, and applications to chemistry.

P. GASSENDI (1592–1655)

Pierre Gassendi, French philosopher and mathematician, was one of the most influential atomists of the period. His works (mainly 1647–1649) cover Greek atomism, Torricelli's work on the vacuum, small microscopic particles, etc. He developed a comprehensive non-mathematical theory of atomism, and suggested that atoms are kept at a distance from one another by actions between them. Gassendi went beyond Democritus in introducing the idea of a definite minimum size for the ultimate atoms.

1650 Thomas Hobbes (1588–1679), English philosopher, was an atomist, but believed that the spaces between hard atoms were filled with fluid.

G. A. Borelli (1608–1679), Italian physicist, developed a kinetic atomic theory, using rational arguments to prove that bodies must consist of a

finite number of separate hard particles in movement, physical properties being ascribed to the
varied shapes of the particles and the action of
external forces pushing them, in the case of gravitation towards the centre of the earth.

During the XVIIth century, partly as a consequence
of the achievements of Galileo and Kepler, there spread through
Europe a new outlook, of which Descartes was a leading ex-
ponent. The scholastic emphasis on substantial forms, essences,
qualitative principles, and the "elementary substances" earth,
fire, water, and air, was gradually displaced by attention to the
"primary" quantitative properties, the spatial arrangements and
motions first of gross matter and later of atoms. This new attitude came to be known as "the mechanical philosophy." With
Bacon, Galileo, and Kepler the scholastic "How?" was replaced
by the scientific "How?," which for Galileo and Kepler meant,
"What are its mathematical laws?"

The effect of this trend on atomic ideas is evidenced in
the fact that between 1646 and 1691 at least seven European
scientists (five of them before Newton) made or discussed estimates, derived from physical measurements, of an upper limit
for the size of the smallest units of the material used: Magnenus
(1646), Charleton (1654), Gassendi (1658), Boyle (1669),
Leeuwenhoek (1680), Newton (various), and Halley (1691).
These estimates were based on experiments with gold leaf,
microscopic particles, incense, smoke, flame, dust, etc., and
most of them yielded upper limits (for the diameter of the
smallest particles of the substance) of around 10^{-4} cm. or
less. They were the earliest quantitative applications of the idea
of "atoms" (smallest units of materials) to measured proper-

ties, and it is no chance that they were made when the new attention to quantity (after 1600) blended with the lively interest in atomism (also from 1600 onwards: Bodin, Galileo, Garleus, Basso, Sennert, Berigard, Borelli, Huygens, as well as the above seven names). Thus the period 1640–1700 marked the birth of *experimental quantitative atomism.* (These experiments gave no positive evidence of a *final limit* to divisibility; nor did Daltonian chemistry; nor can any single experiment.)

The aggressive rise of physical atomism as an adequate explanation of the universe from 1640 onwards provoked a crusade (1660–1700) against it as being anti-Christian. Thomas Cudworth, an English Platonist, argued (1678) that atomism only covers the corporeal world, the incorporeal being evidence of God.

On the other hand, many Christian philosophers and scientists defended atomism. Joseph Glanvill in his *Plus Ultra* (1668) and *Philosophia Pia* (1671) defended the atomic views of some of the founders of the Royal Society against the charge of impiety. He wrote: "How absurd [the Epicurean] philosophy is, in supposing things to have been made and ordered by the casual hits of *Atoms,* in a *mighty Void.* . . . The opinion of the world's being made by a fortuitous jumble of Atoms is impious and abominable." The "Restorers of the Corpuscularian Hypothesis . . . think that the various motions and the figures of the parts of matter are enough for all the Phenomena. . . . But this they suppose and teach: that God created matter and is the Supreme Orderer of its motions. Haphazard atomism is an ungodly idea; the observed order in nature is proof of God's hand." Thus Glanvill believed that atomism was a necessary hypothesis, but that there must be a principle of order underlying the atoms' arrangements and motion.

In 1654 Walter Charleton, London physician, published a valuable survey of atomism. He proved that there were more than 7.10^{17} "atoms" in a grain of frankincense.

1660 R. BOYLE (1623–1691)

Between 1661 and 1666 Boyle clarified the conception of a chemical element, using both mediaeval and atomic ideas. Around 1670 he wrote an essay, "About the Excellency and Grounds of the Mechanical Hypothesis" (here "mechanical" means "atoms in motion"). Generation, alteration, and corruption were, for Boyle, mechanical processes resulting from the motions and rearrangement

1670 of hard atoms, perhaps surrounded by a cloud of heat effluvium, and he looked forward to a universally applicable atomic theory. Heat is atomic vibration. Boyle conceived a complex hierarchy of particles, primary, secondary, etc. As an alchemist, he was convinced of the unity of matter and the transformability of the various atoms. Following Galileo, he distinguished between primary qualities (shape, motion) and secondary (colour, smell, etc.).

R. HOOKE (1635–1703)

Microscopic studies led Hooke to suggest (1665) that regular forms of many kinds, particularly crystals, arise from arrangements of "globular particles." Four such spherical particles compose a regular tetrahedron in crystals of alum. Heat is an oscillatory motion of smaller particles, and all particles are in vibration.

Rohault, French physicist, interpreted (1672) solids as systems in which the corpuscular parts re-

1680 tain their relative positions. Guglielmini (1680) sought to account for crystal cleavage planes by assuming ultimate units, each of crystalline form.

C. HUYGENS (1629–1695)

       Huygens, Dutch mathematician, astronomer, and physicist, conceived the idea of a comprehensive quantitative theory based solely on the kinetics and (pre-Newtonian) mechanics of atomic motions determined by conservation principles. He followed Gassendi and assumed small, perfectly hard particles in rapid movement, all details of their collisions being regarded as irrelevant. This theory, which Huygens applied to gravitation, atmospheric

1690   pressure, light (as waves of particle vibrations), cohesion, etc., was the high point of pre-Newtonian kinetic atomism and was based on what may be regarded today as a sound epistemological basis. Huygens followed Hooke in suggesting (1690) that regular forms arise "from the arrangement of the small invisible and equal particles of which they

1700   are composed," using "little particles, infinitely hard." He published a substantially correct drawing of an atomic model of a crystal.

NEWTON (1642–1727)
Leibniz (1646–1716)

Isaac Newton derived his atomic ideas from the Greeks, Lucretius, Gassendi, and his own teacher, Isaac Barrow. Though in some contexts Newton maintains a neutral position, in others he takes atomism for granted and expresses himself with the utmost clarity, as in the famous passage: "It seems probable to me that God in the Beginning form'd Matter in solid, massy, hard, impenetrable moveable Particles. . . ." (*Opticks,* 1718 edition). In his *Principia* (1685–1687) Newton, by applying the principle of dynamical similarity to particle interactions, showed that Boyle's Law for gases could be derived on the assumption that they consist of hard particles repelling each other inversely as the distance. He considered both attractive and repulsive forces, and supposed that heat was due to the motion of particles. These forces replace the hooks, antlers, and branches of some earlier atomic models. Newton ascribed cohesion either to pressure or to attraction of particles.

In his *Opticks* (1704 edition) Newton obtained an upper limit of about 10^{-5} cm for the size of the smallest soap particles, from estimates of the thickness of soap bubbles based on optical properties.

David Gregory, Oxford astronomer, summarized in his journal a talk he had with Newton (December 21, 1705) on the *hierarchical* atomic structure of matter, the basic units being grouped into composite particles, these again into particles of the next order, and so on (corresponding perhaps to our nucleon, atomic nucleus, molecule, crystal, etc.). Newton preferred this hierarchical structure to a simple lattice structure because he believed it was better adapted to explaining chemical properties, e.g. in terms of one basic particle. The following table is based on Gregory's note of Newton's speculations on the proportions of space occupied by matter and empty space, in systems of different orders, on a simple assumption:

	Occupied space	*Empty space*
Basic particles	Fully occupied	0
Smallest composite particles	$\frac{1}{2}$	$\frac{1}{2}$
Next order	$\frac{1}{4}$	$\frac{3}{4}$
Next	$\frac{1}{8}$	$\frac{7}{8}$
Next	$\frac{1}{16}$	$\frac{15}{16}$
Next	$\frac{1}{32}$	$\frac{31}{32}$

Newton suggested that "Nature accomplishes nearly all the smaller motions of their particles by some other force [i.e., other than gravitation] of attraction and repulsion, which is mutual between any two of these particles." He also proposed a corpuscular theory of light, the motion of the corpuscles being guided by waves.

Newton's prestige was so great, and his concept of force so successful, that his support of atomic ideas and his atomic derivation of Boyle's Law (though incorrect) marked an important stage in the advance of quantitative atomism, though he did not himself develop any general atomic theory. In conversation, Newton said he regarded Gassendi as a fine thinker, whose views on atomism he was proud to share.

G. W. Leibniz, the German philosopher and mathematician, began as a material atomist, but later (1714) developed his non-physical doctrine of *monads* (from μονος, unit), or "simple substances" without extension, shape, position, or movement. They are "metaphysical points; they possess a certain vitality and a kind of perception, and mathematical points are their points of view to express the universe." Though the Leibnizian monads—which are similar to earlier ideas of Bruno, van Helmont, and the English physician Francis Glisson (1596–1677)—are not physical particles, they exerted, through Boscovich, an important influence on the development of mathematical atomism.

1710   The Italian philosopher G. B. Vico (1668–1744) held (1710) that the world is constituted of *point-centres of action*. These are more physical than Leibniz's monads, for they possess location, give rise to extended forms, display tendencies to movement, and do not possess perception. They are halfway between Leibniz's *monads* and Boscovich's *puncta*.

J. Freind, Oxford chemist, developed (1712) a theory of universal attractive forces between hard particles, decreasing rapidly with the distance and greater on certain sides than on others, as the basis of chemical properties. He suggested that the figures of the smallest particles in crystals may be different from the crystal form.

1720
1730

E. Swedenborg (1688–1772), Swedish philosopher and theologian, developed (1734) a doctrine of *natural points* with a tendency to motion, as the origin of all geometrical and mechanical phenomena. He regarded magnets as collections of elementary particles in regular alignment.

1740 D. BERNOULLI (1700–1782)

David Bernoulli, one of the famous Swiss family of mathematicians, taking up the ideas of Huygens and other XVIIth-century atomists, produced (1738) the first post-Newtonian kinetic theory of fluids, and derived Boyle's Law of gases, assuming small elastic spheres.

1750

Around 1750 Benjamin Franklin, the American statesman and scientist, considered that "the electrical matter consists of particles, extremely subtle," which pervade all bodies.

R. J. BOSCOVICH (1711–1787)

R. J. Boscovich, Jesuit mathematician, astronomer, and physicist, published in his *Theoria* (1758) the first general mathematical theory of atomism, based on the ideas of Newton and Leibniz but transforming them so as to provide a *programme for atomic physics*. This used many novel postulates which have since been exploited: (i) There exists one class only of fundamental particles, all identical; (ii) these are permanent quasi-

material point-centres of action, finite numbers of them con-
stituting finite systems; (iii) they obey an oscillatory (alterna-
tively repulsive and attractive) law of interaction, dependent on
the distance between each pair; (iv) this law eliminates the
scale-free similarity property of the Newtonian law, permitting
stable combinations and orbits of definite sizes; (v) this complex law, whose constants are to be determined by subsequent
experiment, is to cover the entire variety of physical and chemical properties; (vi) the theory is relational, only distances between pairs being involved, and all space and time measurements
being in some degree affected by all relative motions; and (vii)
it is kinematic rather than mechanical, mass being introduced
as a pure number determined by counting the number of point-
particles in a system.

Boscovich's theory made no quantitative predictions,
but it exerted a great influence on XIXth- and XXth-century
physics, for it represents the ideal mathematical atomism with
the minimal assumptions (central two-body actions dependent
on powers of the distance). Two features are outstanding: A,
Boscovich eliminated the Democritan-Newtonian dualism of
occupied and empty space and substituted a *monism of spatial
relations;* and B, by doing so, and by using point-centres, he
clarified the meaning of *physical structure:* as a three-dimensional pattern of points. We shall return to these issues in the
next chapter.

Boscovich's point-atoms were no arbitrary invention;
they appeared as a natural idea to many XVIIIth-century minds.
They had been prepared by Leibniz, Vico, and Swedenborg,
and two other thinkers reached the same idea almost simultaneously. John Michell (1724–1793), English scientist, held
that the ultimate particles of matter are point-centres of action,

and developed this in a letter to Priestley. (Michell's fertile ideas were neglected and forgotten; he was too modest for his colleagues to take seriously.) And Immanuel Kant (1724–1803), the German philosopher, at various periods in his life argued that matter does not fill space by absolute impenetrability, but by repulsive forces between ultimate particles localized at points, reaching this idea soon after Boscovich, and perhaps independently.

Experimental quantitative atomism had begun tentatively in the second half of the XVIIth century; theoretical quantitative atomism, initiated by Newton and Bernoulli, became with Boscovich a scheme for a future unified physics, speculative because awaiting the necessary facts.

It may seem surprising that in the middle of the XVIIIth century, when little direct evidence for atomism was available, a thinker such as Boscovich should devote many years to developing a systematic mathematical theory which could at that time be only speculative. Yet this was a natural expression of the situation in 1700–1740. On the one hand atomism had always been a speculation or intellectual exploration from the known large towards the unknown small; from 1600 onwards it became inevitable that this speculative idea should be married to the principle of quantity; and even Newton, in one sense the least speculative of thinkers, whose primary aim was to describe the phenomenon in necessary and sufficient mathematics, considered it worth while showing that Boyle's Law could be derived on certain mathematical assumptions. Thus Boscovich was impelled to his task by three great agencies: the tradition of atomism, the principle of quantity, and the Newtonian concept of force.

1750
1760
1770   Joseph Priestley, English chemist, called attention (1772) to the simultaneous and independent use of point-particles by Boscovich and Michell.
1780   In 1784 R. J. Haüy, French mineralogist, considered that a crystal of any type could be subdivided into correspondingly shaped ultimate solid units, or *molecules integrantes*.

In the same year G. L. Le Sage, French-Swiss mathematician, derived the inverse square law of gravitational attraction, by assuming a random distribution of high-velocity minute particles pressing bodies towards one another and shielding each other.

1790   J. L. Proust (1754–1826) French chemist, proved (1799–1807) that true compounds contain chemical elements in constant proportions. C. A. Coulomb, French physicist, established (1785–1789) his law of the attraction and repulsion of electric charges.

B. Higgins

Bryan Higgins, English chemist (whose nephew W. Higgins anticipated certain of Dalton's ideas) held that chemical combination in definite

1800   proportions, which he called "saturation," is the result of the balancing attractive and repulsive forces between chemical particles.

Count Rumford and Sir Humphrey Davy interpreted heat as motion of molecules. Davy sought to interpret acidity as an expression of the internal structure of molecules, and towards the end of his life expressed his preference for the Boscovichian point-atom.

J. Dalton (1766–1844)

The XVIIIth century appears, in retrospect, as a relatively quiet period in the history of atomism, but the background was being prepared for the dramatic advances of the XIXth. On the empirical side, work was done on stoichiometric relations in chemistry and on the properties of crystals, and in mathematical theory Bernoulli and Boscovich formulated methods whose value became fully evident only much later.

This relative pause ended sharply at the turn of the century, and the discoveries of the first three decades of the XIXth century evidence a new empirical vigour:

1803	Dalton
1811	Avogadro
1812	Wollaston
1815	Young
1816	Berzelius
1827	Brown

To their work, all directly related to observations, must be added Prout's speculation of 1815.

Of all these, Dalton's achievement was the most direct and immediately influential. He formulated the principle that each chemical element is composed of identical atoms—which he imagined as hard and permanent—of fixed characteristic mass, compounds being combinations of atoms of different elements in simple numerical proportions, and he defined "atomic weight" as the mass of an atom relative to that of a hydrogen atom.

The success of this principle in covering many of the stoichiometric facts of simple chemical processes did not "prove the truth of atomism" (in a strict sense this is impossible), but it demonstrated in the most direct possible manner the value of the atomic approach. Indeed Dalton's idea was the first systematic and successful application of atomism to quantitative observations based on special quantitative assumptions regarding classes of material particles—Huygens, Newton, and Boscovich having made only tentative and general assumptions. Moreover, Dalton's principle, with minor modifications, remains today one of the most striking evidences of the particulate structure of matter. But it was not reached as a fortunate "inference" from chemical measurements. Dalton had studied Newton's atomic derivation of Boyle's Law, and thus had inherited the great tradition of atomic ideas in its Newtonian form. Moreover, his first active interest in atomism arose from the possibility of applying it to physical problems such as the properties of gases.

Dalton's unique achievement thus lay in taking an idea from what was at that time *speculative physics* and applying it successfully to *quantitative chemistry*. And he did this in a direct and literal manner: he visualized atoms as hard spheres, and chemical compounds as stable arrangements of these spheres, and so laid the unassailable foundations of *structural chemistry*. Boscovich had anticipated the necessity for such a science; Dalton took the first step towards realizing it. Many chemists during the XIXth century failed to appreciate the importance of this step, and of three-dimensional visual images of chemical processes. As it proved, such images have laid the basis for one of the realms of exact science where facts, once established, can, it seems, never be refuted. Particular molecules have definite mass-relations and stereometric properties, and it seems that no Einstein can ever question the fact.

Though Dalton did not construct three-dimensional models for his compounds, his imagination was visual and concrete, and he was concerned primarily with static arrangements. For example, following Hooke and others, he held that crystal forms are due to the regular arrangements of ultimate particles, and as early as 1803 he made a step towards Bernal's theory of liquids (1958) by treating the structure of water *geometrically,* as stationary assemblies of close-packed spheres.

1800   The Marquis de Laplace developed (1806) a molecular theory of capillarity, using short-range forces.

1810 A. AVOGADRO (1776–1856)

In 1811 Avogadro published his hypothesis that gases at the same temperature and pressure contain equal numbers of molecules in equal volumes.

Wollaston (1812) suggested, following Hooke, that the unit tetrahedrons, octahedrons, and cubes assumed in crystals should be replaced by elementary particles which are perfect spheres (these were later replaced by representative points).

T. YOUNG

Thomas Young, English physician and scientist, inferred (1815) from observations on capillarity, using an atomic theory, that "the diameter or distance of the particles of water" was between one-half and one-tenth of 10^{-9} inches. This is the earliest known determination of the size of molecules or the space occupied by them in a liquid, following the upper limits established in the XVIIth century.

W. PROUT (1785–1850)

Prout, English physician, suggested (1815) that the atomic weights of many elements were integral multiples of that of hydrogen and that all atoms might be composed of hydrogen atoms.

J. J. BERZELIUS (1779–1848)

 Berzelius, Swedish chemist, drew up a table of carefully determined atomic weights (1818), which disproved Prout's hypothesis. He assumed that elements are electrically polarized by the possession of positive and negative parts.

1820 A. L. CAUCHY (1789–1857)

 Cauchy, French mathematician, following Newton, inferred (1820) from optical dispersion that matter is non-homogeneous and non-continuous.

 A. M. Ampère (1775–1836) showed (1825) that small plane currents flowing without resistance in closed curves are equivalent to particles of a magnet, and so developed his conception of molecular currents.

 Robert Brown, English botanist, observed the "Brownian" movement of small particles (1827).

1830

 J. A. B. Dumas, French chemist, wrote (1837): "Atoms are nothing more than molecular groups. If I had the power, I would expunge the word *atom* from the vocabulary of the science, fully persuaded that it goes further than the experiments."

1840 M. FARADAY (1791–1867)

Faraday's role in the history of atomism is of peculiar interest in view of the present particle-field duality. For his work on electrolysis led directly to the conception of units of electricity and stimulated the discovery of electric *particles,* while his discoveries in electromagnetism were used by Maxwell as the basis of his theory of the electromagnetic *field.* Thus both electric particles and electric fields appear to have their main source in the work of one man around 1830–1850.

In 1833–1834 Faraday inferred from his electrolytic measurements that chemical equivalents correspond to equivalent quantities of electricity. He was acquainted with the various atomic models in use at the time, and in 1844 expressed a preference for Boscovich's point-atoms, for "mere centres of force, not particles of matter." But he was a scrupulously cautious thinker, and though his experiments linked electrical phenomena with the tradition of atomism, he did not himself emphasize what seems to us the natural inference: that electrical charge must exist in discrete units. Though electron theories were developed by Fechner and Weber in the 1840's, it was left to Stoney (1874) and Helmholtz (1881) to suggest that Faraday's work *proved* that since matter is atomic, electricity must possess a quasi-material and atomic structure.

We now turn from electrochemistry to electromagnetism. In his studies on induction and diamagnetism involving chiral properties of finite circuits and magnets, Faraday considered that the ideas which provided the simplest and most natural starting point—particles with instantaneous action at a distance—were inadequate, and he found himself (following Oersted, who had conceived an "electric conflict" dispersed through space) picturing the experiments in terms of stationary or vibrating, straight or curved, lines or *discrete tubes of force* normally linking particles of matter, and directly representing the observed effects. Central actions were apparently inadequate because electromagnetic actions involved *curved* lines of force, were dependent on *time rates,* and were sometimes (presumably) propagated, like vibrations along the tubes.

Faraday at one time regarded his discrete vibrating tubes as a substitute for the aether, or continuous medium filling space, though at another period he developed a theory of "polarity" associated with the entire sphere of action of electric and magnetic bodies. A dielectric would then be composed of electrically polarized "contiguous" particles. In 1864–1873 Maxwell adopted this idea of polarity and smoothed out Faraday's discrete tubes into a continuous field. Thus it was not Faraday's observations but the mathematician's preference for a continuous spatial function, and the macroscopic experimental confirmation of that function, that established the conception of the electromagnetic field, into which discreteness had later to be reintroduced by Lorentz and Larmor by modifying Maxwell's work to provide an electron theory.

If they had both lived to experience the direct experimental discovery of the electron (1897), Faraday might have said to Maxwell, "There, I always thought you had slipped in dropping my discrete elements in your continuous field, merely for mathematical convenience in representing large-scale effects."

1850   W. WEBER (1804–1890)

Weber, following Fechner in assuming the corpuscular nature of electric currents, produced (1846) the first "electron theory," with forces dependent on relative velocities as well as distances, and later (1871) a theory of magnetism based on molecular currents composed of an electrical charge moving in an orbit around a fixed charge of opposite sign.

J. P. Joule (1818–1889)

R. J. E. Clausius (1822–1888)

Between 1845 and 1857 (following Bernoulli), Waterston, Joule, Krönig, and Clausius developed the basis of kinetic theory.

In 1851 Joule calculated the mean speed of gas molecules, the first reliable determination of a kinematic molecular magnitude.

Frankland (1852) and Kekulé (1858, 1864) introduced the conception of an (integral) combining power or valency of the atoms of chemical elements, determinable in some cases as the number of hydrogen atoms with which it can combine. Kolbe, Williamson, Odling, and Kekulé used structural formulae displaying the linking of atoms in terms of their valencies.

A. A. de la Rive, French physicist, described (1853) the dominant two-fluid theory of electricity, each composed of particles that repel one another, while the particles of one attract those of another.

The mathematician B. Riemann discussed (1854) the two types of manifold: continuous, in which quantities are compared by measurement, and discontinuous, where they can be counted.

1855

In 1855 G. T. Fechner, German physicist and psychologist, published a detailed analysis of atomic ideas, supporting point-centres.

In 1858 Cannizaro called attention to Avogadro's Hypothesis and clarified the distinction between chemical atoms and molecules.

Clausius (1858) calculated the mean free path of molecules.

1860 J. C. MAXWELL (1831–1879)

Maxwell (1860) and Boltzmann (1868) further developed the kinetic theory of gases, using elastic spherical particles.

L. BOLTZMANN (1844–1906)

Boltzmann wrote on "The Indispensability of Atomism," discussing alternative conceptions.

Maxwell (1864, 1873) replaced Faraday's discrete tubes of force by a smoothed-out continuous function in his mathematical theory of the electromagnetic field.

1865 J. LOSCHMIDT (1821–1895)

>Loschmidt (1865) made the first reliable estimate of the diameter of a gas molecule, following Young's early estimate (1815). Maxwell called attention (1873) to this "fact of tremendous importance." Similar estimates were made by Kelvin (1870) and by Stoney (1888).

D. I. Mendeleeff (1834–1907)

J. Lothar Meyer (1830–1895)

1870
1875

>In 1869 Mendeleeff and Lothar Meyer published their periodic classification of the elements.

>G. Johnstone Stoney inferred (1874) from Faraday's work on electrochemistry the existence of a natural unit of electricity and estimated its value.

>In the same year Le Bel and van't Hoff associated physical properties with the spatial arrangements of atoms in the theory of the asymmetrical carbon atom, thus creating the science of stereochemistry.

>In 1875 Maxwell gave the probable value of the maximum diameter of a hydrogen molecule as about 5.10^{-8} cm, and recognized that the values of specific heats showed that there could not be equipartition of energy between the degrees of freedom of molecules.

The many successes of atomic theory during the early and mid-XIXth century produced a lively discussion of its implications around 1870–1880, and contrary views were expressed with great vigour. Du Bois–Reymond represented the classical mechanical attitude when he wrote: "The resolution of the changes of the material world into motions of atoms caused by their central forces would be the completion of natural science." It is instructive, in retrospect, to see how many physicists maintained with intense conviction the orthodox doctrine of central forces at the very moment when the theoretical importance of chiral electrical actions was being increasingly recognized. Though such basic issues as the relation of electricity to matter and of discontinuity to continuity were still entirely obscure, many leading physicists were dogmatic regarding the fundamental character of the classical concepts of space, time, and mass. This situation, to us paradoxical, was possible because at that time no comprehensive physical theory was yet in sight, and even the conception of a unified theory—as we understand it today—had hardly begun to influence the thinking of physicists, though ideas of the unity of natural phenomena and of the conservation of all energy had increasingly influenced physical thought after 1830–1850. It was still appropriate for physicists to think in compartments.

One of the most striking facts in the history of fundamental physics is that for a century, from Fresnel (1821) to de Broglie (1924), the particle theory of matter and the wave theory of light coexisted without provoking physicists to discover some physical relation between these contrasted ideas, as though they had coexisted in independence or pre-established harmony from some original act of creation. Mathematical analogies and connections were worked out by many, from Hamilton's correlation of dynamical trajectories with optical rays to the electron theory treatments of periodic waves emitted by orbital electrons (in spite of the loss of energy!). But no fundamentally novel observable consequences were drawn from any such unifying principle or analogy until 1923. It seems that in spite of the persisting ideal of unification (one primary type of matter, one form of force of which particular forces were expressions, and so on) particles and waves were unconsciously regarded as creatures of different species incapable of a fertile marriage. The idea of a genuinely unified physics did not become practical politics for serious minds until after the successes of electron theory and, even more important, after the undermining of all classical categories by relativity and quantum theory.

This is shown most clearly by Maxwell's situation. For though he made fundamental contributions in relation both to discrete matter (kinetic theory) and to continuous electricity (electromagnetic theory) he appears to have given up an early attempt (1855) "to construct two theories, mathematically identical, in one of which the elementary conceptions shall be about fluid particles attracting at a distance while in the other nothing [mathematical] is considered but various states of polarisation, tension, etc., existing at various parts of space." Maxwell died in 1879, just before the problems of the relations of matter and electricity, of material discontinuity and apparent electrical continuity, became topical. Maxwell's early hope was justified seventy years later, but with a difference: quantum mechanics created a dual theory with mathematically complementary (not identical) branches treating particles and fields.

In a paper on "Molecular Constitution of Bodies," written in 1875, Maxwell expressed the classical view with great force: "We can not conceive any further explanation to be either necessary, desirable, or possible, for as soon as we know what is meant by *configuration, mass,* and *force,* we see that the ideas which they represent are so elementary that they can not be explained by means of anything else." Yet at that time Maxwell had no idea how discrete atoms with mass could be linked with his own continuous electromagnetic equations, though he had noted in his *Electricity and Magnetism* (1873) that the assumption of "one molecule of electricity" is "out of harmony with the rest of the treatise."

This illustration of the somewhat casual separation of two aspects of physical theory in Maxwell's mind can be supported by others. For example, in his brilliantly concise statement of dynamical principles, *Matter and Motion* (1877), he dismissed both molecular and electrical properties with the briefest of passing references, as though irrelevant in considering the energy of a physical system. Though aware that all knowledge of space is relative, Maxwell still believed that absolute space is necessary in dynamics: "The position seems to be that our knowledge is relative, but needs definite [i.e., absolute] space and time for its coherent expression." That phrase "seems to be," so unlike his positive clarity elsewhere in this essay, indicates that Maxwell dimly sensed the contradiction that opened the door to the theory of relativity.

If such separate compartments could coexist in a mind of the highest order, it is not surprising that opposite views on atomism were being held simultaneously by philosophers, physicists, and chemists of different schools. Many leading chemists, for example, refused to take atomic interpretations literally until the end of the century, as did also several philosophers of physics. Stallo protested (1881) that the concept of atoms arose merely from "the re-ification of the concept of cause," and Mach, who as a younger man had accepted physical atomism, came to regard atoms as hypothetical inferences convenient for the economical symbolizing of experience—these attitudes reflecting earlier philosophical critiques of language. Though these sceptical views had no effect on the triumphant progress of experimental and mathematical atomism, they may be relevant now that the concept of a "physical particle" has been radically transformed and its meaning has become less clear.

In the meantime, from 1860 to 1890, theoretical physicists were searching for mechanical models of the electromagnetic medium, in order to explain both its properties as a continuum and (in a few minds, as a secondary issue) the existence of discrete localisable entities, such as the atoms and molecules of chemical and kinetic theory. This theoretical exploration was perhaps the most bizarre in the history of physics; it led to Helmholtz's vortex rings (1858), Maxwell's vortices and rolling particles (1861–1862), Kelvin's vortex atom (1867), vortex sponges, and aether squirts, and Fitzgerald's wheels linked by rubber bands (1885). Only later, after 1899, did it become clear that such models of the properties of a physical medium, supposed to vary in its state from point to point, could not be compatible with the relational character of all observations.

In the meantime, from 1880 onwards, the view that *electrical* properties were theoretically primary steadily gained ground, material atomism yielding place to electrical atomism and inertia being treated as an electrical property. Though it was not realized at the time, this change from signless material properties to dual (positive and negative) electrical ones implied, as Eddington pointed out in 1939, a shift from the analysis of *substance* to that of *form* (polarization, structure).

It is important to realize that even at the height of classical mechanical atomism it was evident to some scientists that chemical atoms might not be unchangeable units but might be composed of simpler ones, as indeed Prout had suggested in 1815. During the middle decades of the century considerable attention had been paid to the contrast between the supposed unity of matter and the actual variety of the chemical elements.

In the 1870's, after the discovery of the periodic table, Kelvin and Spencer emphasized that chemical atoms were not necessarily indivisible or eternal, and might be merely arrangements of different numbers of uniform ultimate units. But until 1896 there was no reason to doubt the natural stability of atoms.

H. v. HELMHOLTZ (1821–1894)

Helmholtz, following Stoney, showed (1881) that Faraday's electrolytic experiments implied the atomicity of electricity, and explained integral valency in terms of positive and negative units of electric charge.

J. J. THOMSON (1856–1940)

Thomson invented the concept of "electric mass" (1881).

Stallo observed (1881) "that there is hardly a treatise on modern physics in which atoms or molecules are not compared to planetary or celestial systems."

1885

In 1884 Arrhenius published his theory of the ionisation of dissolved electrolytes.

In 1886 W. Crookes predicted the existence of isotopes, and suggested that all atomic properties depend on the existence of an ultimate unit of electric charge.

1890

Between 1885 and 1894 three mathematicians (Federow, 1885; Schoenflies, 1891; and Barlow, 1894) independently proved (subject to minor errors) that 230 distinct types of crystal structure (i.e., 230 different space-groups determining the homogeneous partitioning of three-dimensional space) are possible, on the assumption that crystals are linear arrays of equivalent "molecules," each equivalently placed (at a point) with respect to its neighbours.

In 1888 J. J. Balmer discovered his formula for the wave lengths of the hydrogen spectrum.

In 1891 Johnstone Stoney gave the name *electron* (*e*) to the natural particle and unit of electricity to which he had drawn attention in 1874, linked atomic electrons with optical properties and chemical valency, and suggested the possibility of a magnetic unit of angular momentum.

1895   Lorentz (1892) and Larmor (1894) developed electron theory by introducing atomic ideas into Maxwell's electromagnetic theory. Larmor inferred (1894–1904) from experiments on rotating electrical systems that electric currents must consist of discrete charges.

Direct Experimental Discovery of the Electron. In 1897 Thomson, Wiechert, and others determined the charge-mass ratio for cathode-ray particles, identifying these with electric "atoms" (electrons). Lorentz explained the Zeeman effect (discovered 1896), using electron orbits.

In 1896 Becquerel discovered the spontaneous radioactivity of uranium, in 1898 Pierre and Marie Curie found the same property in radium and polonium, and in 1902 Rutherford and Soddy proposed the transformation hypothesis.

Rayleigh (1899) identified monomolecular surface films, basing a theory of them on the assumption of smooth solid spheres.

1900   Rydberg (1900) and Ritz (1908) formulated the combination principle for line spectra.

The discovery of radioactivity (1896) and of the electron (1897) marked the culmination of a century of dramatic empirical advances in the realm of atomism. For many decades evidence in favour of atomism had steadily been accumulating from chemistry, electrochemistry, stereochemistry, optics, heat, crystallography, and finally from discharges through gases and the disintegration of the chemical atom. Until 1880–1890 classical atomic concepts had prevailed; from 1890 to 1900 electrical concepts began to take over. The idea that discrete electric charges constitute the basis of electromagnetic phenomena was being vigorously exploited by Helmholtz, Thomson, Arrhenius, Larmor, and Lorentz. There is irony in the fact that just when this essentially classical concept of an ultimate electrical particle appeared to be finally justified by its discovery during a few years of intense experiments (1896–1898), all classical ideas were about to be undermined and gradually replaced by the new methods of relativity and quantum theory.

As the new century opened, three contrasted dynamical doctrines held the field, based respectively on (1) a continuous physical medium capable of activity (implying "absolute space"); (2) hard finite atoms undergoing collisions; and (3) point-centres of force with relational action at a distance. But none of these solved the problem of the relation of continuity to discontinuity in a manner compatible with experiment.

The end of classical physics and the opening of XXth-century relativity and quantum physics may be interpreted as a tremendous, partly unconscious, shifting of the continuity-discontinuity axis. In one sense, both the discrete and the continuous aspects of physics were gaining ground because new realms were being discovered. But this movement took place in two distinct operations which have not yet been fully coordinated.

1. *Relativity Theory*. Boscovich (1758) and Mach (1883) had stressed the relational character of all spatial observations and the consequent relativity of inertial and mechan-

ical properties. But continuity and contact action implied, for the classical mind, a physical medium throughout space in a particular condition at each point, and this medium fixed in absolute space was not given up easily. In 1899 Poincaré asserted that absolute motion is in principle indetectable, and in 1904 he formulated his "principle of relativity," which extended to electromagnetic phenomena the mechanical relativity of Boscovich and Mach. The resulting mathematical transformations were formulated by Lorentz (1899), improved by Larmor (1900–1903), and called by Poincaré (1905) the group of Lorentz transformations. The new outlook was crowned by Einstein in his Special Theory (1905) based on the postulate of the universal constancy of the velocity of light. Minkowski introduced the elegant four-dimensional representation in 1907. Klein suggested (1910) that "relativity" should be replaced by "invariance relative to a group."

This radical transformation of physical ideas rested on two primary conceptions: the relational character of space, and the universal invariance of c, the velocity of light. The continuity of a physical medium filling space was replaced by the continuity of the changing spatial relations of material entities expressed in a relativistic geometry, first Euclidean and later, in the General Theory (1916), non-Euclidean.

2. *Quantum Theory*. Meantime Planck had created the new realm of quantum physics by his discovery of the quantum constant h (1900). The discreteness of e implied an atomicity in the factor responsible for electrical forces, and hence could be accounted for by the existence of positive and negative electric *particles*. But the discreteness of h could not be traced to any atomicity of localised quasi-material *entities*, for it implied

a discreteness of *process,* of action or angular momentum. Discontinuity was gaining ground, but in a new realm.

This discrete quantization of process was distressing to minds trained in the XIXth century before the discovery of h. Planck (b. 1858) and Einstein (b. 1879) both resisted it, hoping to the end to eliminate it in a deeper theory; Bohr (b. 1885) was the first with the courage to take it seriously and to create a new quantum philosophy. But his Correspondence Principle required a continuing reliance on classical ideas as the basis for the advance into the quantum realm. And none of Bohr's followers has yet shown where a new start should be made: whether in an abstract formalism brought step by step towards classical observations, or in a novel interpretation of all observations. Bohr built the bridge, but the great army that followed him over has failed as yet to establish a stable, independent base on the further bank.

The invariance of c demanded an invariant field representation of Maxwell's equations. To Einstein, Bohr, and Heisenberg the discreteness of e and of h implied the necessity of a deeper-lying unified field equation, contrasted solutions of which would account for this dual discreteness of charge and angular momentum. But half a century of searching has brought no success.

The discovery of h was so surprising, and it conflicted so violently with what had been regarded throughout the XIXth century as basic physical principles, that it has been regarded by some as a perfect example of "pure induction from observations," the facts compelling the scientist's mind to adopt conclusions which could only be reached, and were in fact reached, in this manner. At a superficial level and in a vague sense this view may be correct. But it is historically and psychologically misleading. Ideas do not develop because they are necessary logical inferences from "facts" uncoloured by hypotheses. Before Planck a great tradition, from the Greeks to Boscovich, had speculated regarding a discontinuity or stepped character in the rhythms of process. In Planck's unconscious mental processes in 1900, as he consciously thought "I will try the effect of discrete packets of energy, say proportional to the frequency," a factor was at work which was *either* the direct or indirect influence of this tradition, *or* the same formative process which had created this tradition operating again in him, or a combination of these.

Planck's reasoning was certainly logical at a certain level and was empirically justified, but it was not a logically necessary inference from the facts. For a school of field theorists considers that the discreteness of h will *not* be retained as a primary feature of a theory which correlates e, h, and c (and perhaps other atomic constants) by deriving them from simpler field principles. Thus the continuity-discontinuity issue remains in 1960 fundamentally as obscure as it was to the early Greek thinkers. Beneath the tremendous empirical and mathematical elaborations, the underlying contradiction between these two modes has persisted, though modified by a century and a half of atomic researches and a new statistical interpretation.

It is interesting to note that at the end of the XIXth century and in the first decades of the XXth the idea, later explored in quantum theory, that large-scale determinateness was compatible with micro-indeterminacy began to appear in physical discussions. Two examples: Larmor (1900, written 1898) suggested that ordinary observations were *averages* of indeterminable atomic phenomena, and Exner (1919) that complete randomness of atomic processes underlay the observed regularities.

1900   M. PLANCK (1858–1947)

Planck introduced (1900) his constant of action, h, and the idea of discrete parcels of energy, to account for his formula for black-body radiation.

1901   Kaufmann established (1901–1902) the variation of electron mass with velocity.

J. W. Gibbs in his *Statistical Mechanics* (1901) avoided the use of particular atomic or molecular models, relying on the general laws of mechanics, and introduced the conception of

1902   phase-space, taken up by Planck in 1906. In earlier work Gibbs had exploited the complementarity of position and momentum, later developed in quantum theory.

1903   Lenard inferred (1903) from electron absorption experiments that the atom is mainly empty space, and Nagaoka (1904) speculated on a "nuclear" model atom. Various atomic models were in use: negative stationary electrons in a

1904   diffuse sphere of positive electricity (Thomson, 1903), small electric doublets distributed through the volume of the atom (Lenard, 1903), a small positive nucleus with orbital electrons (Nagaoka;

1905   Rutherford, 1911; Nicholson, 1912; Bohr, 1913), and some others. In 1905 Thomson initiated the systematic study of atomic structure, seeking to account for the periodic table by using planar rings

1906   of orbital electrons in a diffuse sphere of positive electricity.

A. EINSTEIN (1879–1955)

In 1905 Einstein used the conception of localised parcels of electromagnetic radiation (later

1907   called *photons*) with $E = h\nu$, to account for the photoelectric effect.

Nernst (1906) interpreted the zero point of

temperature as a zero point of disorder, rather than of motion or energy.

Einstein's kinetic theory of the Brownian Movement (1904), developed by Smoluchowski (1906), led to Perrin's studies (1908–1909) including the determination of the number of atoms in a gram-molecule.

1908

Larmor showed (1908) that the inertia of positive fundamental particles, being much greater than that of electrons, cannot be entirely electrical in origin.

1909

Einstein noted (1909) the challenge which the dimensionless group (e^2/hc) presents to an atomic theory of electricity.

1910

F. Soddy discovered (1910) and named *isotopes*.

1911 E. RUTHERFORD (1871–1937)

Rutherford measured α particle scattering and proposed his atomic model of planetary electrons round a small positive nucleus (1911), leading to the distinction between *mass number* and *atomic number,* identified with the nuclear charge by van den Broek and H. G. J. Moseley (1913).

Superconductivity was discovered (1911) by Kammerlingh Onnes. Dunoyer (1911) investigated the rectilinear paths of molecules in vacuo.

1912

C. T. R. Wilson made the paths of charged particles visible in his cloud chamber (1912). Von Laue discovered (1912) the X-ray diffraction patterns produced by crystals.

R. A. Millikan measured the electronic charge on oil droplets (1908–1913).

N. BOHR

J. W. Nicholson (1912) suggested that the angular momentum of an atom can only vary by

1913

units of $h/2\pi$ and that line spectra result from this fact. Ehrenfest proposed (1913) that all angular momentum must be integral multiples of h. Finally Bohr (1913) used h to provide a theory of Rydberg's constant and the hydrogen spectrum with

1914
1915

radiative transitions between stationary orbits characterized by quantum numbers, and later developed his Correspondence Principle. Kossel applied Bohr's theory to X-ray spectra.

Wilson and Sommerfeld independently introduced elliptical electronic orbits, and Sommerfeld

1916

(1916) used a relativity correction to explain the fine structure of optical spectra in terms of the

1917
1918
1919

fine-structure constant $(\alpha = 2\pi e^2/hc)$.

Rutherford achieved (1919) the first deliberate disintegration of an atomic nucleus.

Ehrenhaft's work (1909–1937), on an electric charge less than e, remained unconfirmed.

Aston used (1919) his mass spectrograph to determine atomic masses.

1920

Stern made (1920) the first direct determination of molecular velocities.

Rutherford named the *proton* (1920) and Pauli the *Bohr magneton.*

1921

Between 1922 and 1924 Bohr and others developed the closed-shell theory of atomic structure

1922

underlying the periodic table, and in 1924 Pauli formulated his Exclusion Principle.

A. H. Compton used (1923) the Einstein relation, $E = h\nu$, to account for X-ray scattering (Compton effect).

1923 L. DE BROGLIE

L. de Broglie introduced (1923–1925) the idea of *wave-particles,* exploiting the parallel be-

1924   tween corpuscle dynamics and wave propagation, in order to cover the wave and the particle properties of light, and Elsasser (1925) showed that certain electron reflection maxima might be due to De Broglie waves.

In 1924 Bohr, Kramers, and Slater suggested that the conservation principles of energy and momentum might be only statistically valid.

We have now reached the uniquely fertile years of the creation of quantum mechanics. From 1924, when de Broglie's idea began to receive attention, to Dirac's relativistic wave equation of 1929, a rapid sequence of fundamental advances was achieved, unprecedented in the history of physics. During these five years some thirteen theoretical physicists (many of them very young) contributed unquestionably authentic fundamental discoveries which fitted together to complete the brilliantly successful theory of quantum mechanics. For those who experienced them, the years 1925, 1926, and 1927 were incomparably exciting; some three or four entirely new and valid insights into fundamental properties were achieved in each year. Moreover, the result was a unified and comprehensive "fundamental" theory.

It was therefore not surprising that many of those who took part in this dazzling achievement, and also many talented commentators, believed—say between 1928 and 1940—that theoretical physics had been brought to a triumphant conclusion: the formulation of a theory capable of predicting the results of all conceivable experiments.

This claim was either misleadingly expressed or absurd —the result of temporary blinding by excess of illumination *on certain aspects only*. The most distinguished exception, who at once (1930) published a warning, was Heisenberg. He pointed out that in addition to the modifications of classical physics required by the existence of the velocity of light and of Planck's constant, other limitations must be expected to appear connected with the other universal constants. Moreover, Einstein had already made it clear, to those who cared to take heed, that a unified derivation of Maxwell's equations, electron theory, and early quantum theory would have to account for the value of the ratio (e^2/hc). These warning hints were neglected in most treatments of quantum mechanics, and little mention was made of the fact that the empirical validity of the theory had only been proved to a limited accuracy over a restricted range of variables (distance, temperature, energy, etc.).

In retrospect these years invite two reflections:

1. The unexpected and convergent character of the multiple advances of 1924–1929 give unassailable proof of the objectivity of physics.

2. The intimidating effect of this unprecedented success has had a curious consequence: no leading theoretical physicist has yet published a comprehensive analysis, or even a list, of the fundamental problems on which quantum mechanics, in its canonical form, does *not* throw light.

1924 M. BORN
In 1924 Born treated electron orbital perturbations by his new method of quantum mechanics, and Bose, supported by Einstein, developed a new quantum statistics for certain particles, contrasting
1925 E. SCHRÖDINGER (1887–1961)
with the Fermi-Dirac statistics (1926) for others. In 1925 Born recognised in Heisenberg's (1925)

W. HEISENBERG, W. PAULI (1900–1958)

1926   noncommutative multiplication the law of *matrix* multiplication; Pauli formulated his Exclusion Principle; Born, Heisenberg, and Jordan developed the matrix calculus and general transformation theory;

P. A. M. DIRAC

1927   and Dirac his q-numbers based on *Poisson-brackets.* In 1926 Born and Wiener published their *operator* theory, and in 1927 Heisenberg formulated his *uncertainty relations.*

1928   Meantime Schrödinger, de Broglie, Klein, and Gordon, developing de Broglie's idea of 1923, simultaneously (1925) obtained the *Schrödinger wave equation,* and Schrödinger developed his *wave mechanics* and theory of eigen-values, which, fol-

1929   lowing Born's interpretation (1926) of the wave function as a measure of probability, was subsumed into the general transformation theory of quantum

E. P. WIGNER

variables, employing group theoretical methods introduced by Wigner and others (1927–1930). The first definite experimental confirmations of the de Broglie–Schrödinger electron waves were obtained by Davison and Germer, and by G. P. Thomson, in 1927. In 1928 Jordan and Wigner introduced

1930   second quantisation. Between 1929 and 1933 quantum electrodynamics and the general theory of quantized fields was matured (Fermi, Jordan, Heisenberg, Pauli, Wentzl, Rosenfeld).

In 1928–1929 Dirac linked quantum theory and relativity in his linear wave equation for the electron, which stimulated Eddington's attempt at a "fundamental theory" (1928–1944) of the pure numbers, and influenced subsequent work in relativistic quantum theory, including Heisenberg's work on non-linear field theory (1945 onwards). Around 1930 it was believed that only two material particles existed: the *proton* and the *electron*; and one radiative particle, the *photon*.

1931   Wigner classified quantised states of atoms as terms in a power series in the fine-structure constant.

1932   C. D. Anderson discovered the *positron* (1932) predicted by Dirac in 1929, and J. Chadwick the *neutron* (1932), and the science of "nuclear chemistry" was opened.

1933   Heisenberg (1932) developed a theory of the structure of the nucleus, assumed to contain only protons and neutrons.

1934   Fermi developed (1934) the *neutrino* theory of β decay.

1935   Yukawa predicted (1935) what was later called the *meson,* interpreting particles as quanta of force fields.

1936   Anderson discovered (1936) the first meson.

1937   Bohr initiated the "liquid drop" model of atomic nuclei.

1938

Born and J. A. Wheeler published (1939) a theory of fission, and Hahn and Joliot-Curie discovered (1939) fission of uranium.

1939   Several physicists reached the conception of a *chain reaction* (1939), and Fermi, assisted by a unique group of experimental and theoretical workers, produced the first nuclear reactor (1942).

N.B.: No attempt is made here to cover all the significant experimental discoveries made after 1940 in nuclear and high-energy particle physics. Only some of the more important advances affecting fundamental particle theory are included.

1940
1941

1942
1943
1944   In the decade after the end of the Second
1945   World War attention was paid to the classification
       of particles, and the conception of *baryons* (heavy
       particles) and *leptons* (light particles) was grad-
       ually developed.
1946   Powell, Lattes, and Occhialini discovered
       (1946) new types of mesons.
1947   Schrödinger and Tamonaga published (1947)
       an improved quantum electrodynamics interpreting
       Lamb and Retherford's experimental discovery
1948   (1947) of a correction to Dirac's expression for
       the fine structure of the hydrogen spectrum.
          Goeppert-Mayer (1948) established the *magic
       numbers* determining the most stable nuclei.
1949   In 1949 several workers (Schwinger, Feyn-
       man, and others) clarified the idea of mass *renor-
       malization* implicit in earlier work, and in the same
       year Feynman introduced his *graphs* of particle
       transformations.
1950   After 1950 various speculative ideas extend-
       ing quantum mechanics were proposed: Wheeler's
       geons (gravitational particles), Bohm's critique of
       the basis of the theory, etc.
          Fröhlich and Bardeen, using the concept of
       electron-lattice interactions, founded (1950) the
       quantum mechanical theory of superconductivity.
1951   From 1945 onwards, studies were begun of
       the properties of *plasma,* the high-energy elec-
       trically neutral system in which all nuclei are fully
1952   ionised and all electrons free.
          From 1953 onwards Heisenberg investigated
       the possibility of a unified theory of particles based
       on a non-linear spinor field equation.

1953	Gell-Mann and Nishijima proposed (1953) the "strangeness" quantum number.
1954	During the 1950's the theoretical treatment of many-body problems (complex atomic systems) was assisted by the identification of "collective parameters" and "quasi-particles" corresponding to modes of vibration or deformation of a system as a whole.
1955	Segré and others produced (1955) protons and anti-protons.
1956	The neutrino was observed (1956).
	Discussions in 1956 led Lee and Yang to suggest the possible failure of parity conservation in certain weak particle interactions, and Wu and
1957	others confirmed this (1957).
	Improvements were made (1957) in the theory of superconductivity.
1958	Bernal opened the statistical geometrical theory of stationary irregular molecular arrangements in liquids (1958).

CHAPTER 5

Analysis

Wᴴᴬᵀ stands out when we survey this history?

1. *The persistence of the main idea and the variety of its forms and applications.*

2. *The fact that there have been, so far, three primary conceptions of atomic particles.*

3. *The application, from the XVIIth century onwards, of quantitative methods drawn mainly from geometry to an idea which originally lacked quantitative definition, though related to space.*

4. *The remarkable increase in quantitative precision during the last three centuries:*

1640–1680	Many rough estimates for upper limit to size of molecule.
1805–1815	Estimate of diameter of molecule.
1865	Diameter of gas molecule, to two significant places (s.p.).
1897	e determined to 1 s.p.
1911	e to 3 s.p.
1957	e to 4 or 5 s.p.
1960	Nuclear frequencies to 8 or 10 s.p.

5. *The absence, as yet, of any fully satisfactory mathematical correlation of the discrete (particle) and continuous (field) aspects of observations.*

Relativistic corrections in (v/c) are necessary to link the velocities (v) of discrete particles with the continuous field velocity (c). When applied to fields, these corrections eliminate dependence on a physical medium and relate field phenomena to particle velocities; when applied to particles, they relate the particle velocities to field phenomena. Thus the terms in (v/c) link discrete particles to continuous fields, and vice versa. But this recognition has not led to any improved formula uniting continuity and discontinuity in a more elegant and fertile manner.

6. *A movement, evident from 1700 onwards, away from the Democritan-Newtonian dualism of occupied and empty space towards a monism of structural relations, still awaiting definitive formulation.*

7. *The tendency of atomic physics to start by emphasizing the properties of the units, as displayed in simple systems or two-body interactions, and to postpone the study of their ordering in complex systems.*

Until this century, fundamental atomic physics was mainly concerned with simple one- or two-body systems. Two classes of complex systems had already been treated: fully ordered static crystal structures and fully disordered statistical aggregates; but these were not regarded as fundamental. In the first decades of the century greater attention began to be paid to complex dynamical systems which are partly ordered (nuclei, atoms, and molecules, in excited states). XXth-century atomic physics, in many of its applications, is mainly concerned with partly ordered complex systems.

However, no approach to a general theory of the transformations of complex, partly ordered, systems has yet been attempted. Some forms of stable order are already understood:

atomic shells, simpler chemical compounds, and the 230 types of crystal structure. But others, such as nuclei, are not. The ordering of transformations in living cells, organisms, and brains is still obscure. On the other hand research is now more evenly balanced between the study of single ultimate particles and their forms of ordering in systems, and important advances in the latter field are likely in the coming period.

8. *The history of atomism displays two contrary tendencies as yet unresolved:*

(i) To ascribe an increasing number of properties and parameters to ultimate particles: mass, electric charge, spin, life-time, isobaric spin, parity, strangeness, etc. This has been the dominant trend since 1800.

(ii) To maintain the ultimate aim of *reducing* the number of theoretically arbitrary parameters ascribed to particles, as would be necessary in any unified theory of particles. This aim has been expressed frequently, for example by Prout, Einstein, and Eddington; and by others who have sought to establish a relativistic or relational physics in which physical parameters are only ascribed to the spatial relations of particles.

9. *The subtle character of the transformations which the conceptions of particles has undergone:*

From *inert massy units,* arbitrarily associated with actions at a distance, to positive and negative *units of electric charge,* or units of polar interactions with theoretically arbitrary masses. Units of $\pm$ charge imply polarization, or polar vectors which tend to decrease and stable complex systems which tend towards electrical neutrality. This tendency towards neutrality which pervades the electrical realm is treated as secondary, or even as theoretically trivial, in wave-particle theories based on reversibility, symmetry, and invariance.

10. *The continuity and cumulative character of the story, which, like that of science as a whole, is one of novelty within continuity.*

Though there have been several profound changes, and though logically incompatible ideas have sometimes been necessary to cover complementary aspects, the story displays a remarkable sequence of definite conceptual steps, leading collectively towards deeper understanding of structure. Individual discoveries may have been relatively haphazard, but the total effect suggests a grand design which is slowly being revealed as the result of the interplay of developing techniques and steadily improving ideas. This design may perhaps be described, in our current terms, as the interplay of order and disorder, but some of its fundamental features are still obscure.

On the theoretical side the stepped character of the advance is particularly clear, though such steps can be isolated in different ways according to the purpose in view. The following analysis is intended to suggest that some twenty-one steps can be identified which appear necessary and sufficient to link the basic idea of Greek atomism with the fundamental conceptions of atomic theory in 1960. Only some of the earlier or more important names are cited.

Necessary and sufficient theoretical steps leading from the basic conceptions of Greek atomism to those of XXth-century atomic theory.

1. *Duality of Space*	Democritus
There exist two kinds of space: occupied (matter) and empty (void).	
2. *Discreteness of Matter*	Democritus
Ultimate indivisible units of matter exist, with unchanging properties.	
3. *Hierarchical Structure*	Sennert, Basso,
These units are grouped into systems of various orders.	Newton, Boscovich
4. *Properties*	Various
They possess quantitative properties which can be determined by observations.	
5. *Forces*	Newton
In particular, they interact in pairs by forces.	
6. *Equivalence*	Boscovich, Prout
In an ideal theory, all primary particles are equivalent.	
7. *Point-centres*	Boscovich
The primary particles are point-centres of interaction.	
8. *Relational*	Boscovich, Mach, Poincaré, Einstein
Only the changing spatial relations of particles enter basic theory.	
9. *Stability*	Newton, Boscovich, Bryan Higgins
Stable composite systems result from the balancing of forces.	
10. *Chemical*	Dalton
Atoms of chemical elements have characteristic weights; compounds contain these in simple proportions.	

11. Electrical
There exist indivisible units or particles of positive and negative electric charge, linked by a tendency towards neutrality.

Helmholtz,
Stoney,
J. J. Thomson

12. Electron Theory
These electrical particles play a primary role in physical theory.

Weber, Lorentz,
Larmor,
J. J. Thomson

13. Electron/Proton
There is a basic non-equivalence of mass between the positive and negative particles (in low-energy systems).

Larmor,
Rutherford

14. Quanta
h determines discrete quanta of action and unique quantised states.

Planck, Bohr

15. Nucleus
The atom is a central massive positive charge with light planetary negative electrons.

Rutherford

16. Wave-particles
Wave-field representations of particles, which become quanta of fields.

De Broglie,
Schrödinger

17. Statistical
The wave-particle formalism represents probabilities of measurements, subject to complementary indeterminacy of conjugate magnitudes.

Born,
Heisenberg

18. Complete Description
A satisfactory quantum treatment of any particle process must take into account the entire system (initial and boundary conditions, apparatus, mode of observation) within which it occurs.

Bohr

19. Multiplicity	Various

The current interpretation of nuclear and high-energy observations implies the presence of many kinds of elementary particles with varied properties (mass, charge, spin, invariance, modes of n-body interaction, etc.), associated in complex systems with a tendency towards electrical neutrality, balancing of spins etc., and other collective properties.

20. Relativity	Einstein,
Relativistic representations are necessary.	Sommerfeld, Dirac
21. Ideal Theory	Boscovich,
An ideal atomic theory would unify all known effects under a single law.	Prout, Einstein, Eddington, Heisenberg, and many others

Further links will undoubtedly be added to this great chain of ideas. Since around 1950 various new methods are being tried out on particular problems and discussed by specialist groups. None of these has yet gained the status of an authentic contribution to fundamental theory. But a few of them may be mentioned here as promising growing points already visible in 1960:

Collective parameters. Some complex quantum mechanical systems have been shown to possess collective parameters or coordinates representing, either accurately or approximately, excitations of the system as a whole. When such parameters can be identified, they greatly simplify the treatment of complex phenomena. They are conveniently regarded as parameters of new classes of *quasi-particles* (excitons, phonons, rotons, etc.), each representing an elementary mode of motion or deformation of some physical structure. Though no general procedure for identifying appropriate collective parameters has yet (1959–1960) been established, this method of covering the properties of complex systems is attractive, and it may provide a substitute for the more difficult methods of non-linear field theory.

Classification of elementary particles. Since 1950 a consistent classification of particles has been developed, based mainly on their invariance properties. This is of great importance, but it tends to neglect other aspects: mass ratios, and the fact that particles which are formally similar may none the less possess a contrasted theoretical status. For example, it appears that *nucleons* may enjoy a prior theoretical status because they play the major part in determining the initial conditions of atomic processes (apparatus, source of energy, etc.). This suggests that if the particle laws were fully known, all particle processes could be correlated to changing patterns of nucleons.

Principles of invariance and their limits. All known particle laws (excluding the values of the dimensional constants) can be expressed as principles of invariance, covering isolable features of the phenomenon—or in a few cases as the absence of invariance. However, no principle of invariance can be absolute, i.e., accurate to any desired degree, since finite disturbing influences from the environment are always present (radiation, non-homogeneous distribution of mass in the universe, etc.). In effect, physical laws express criteria implicitly used in treating systems as isolable. Theory might therefore benefit from an explicit analysis of the conditions involved in treating any system as if the rest of the universe did not exist, showing how and when and what we can know about any one system, without knowing all about everything.

These are a few of the more promising growing points of current theory. Less attractive lines of research are: the use of non-linear fields, non-local fields, new types of non-Euclidean geometry in very small regions around points, more abstract representations, and the relation of particle fields to gravitation. But every direction must be explored, for a new and widely applicable method may spring into existence unexpectedly out of the most unlikely problem. While attention is consciously directed to some special but intriguing problem, the unconscious preparatory processes of the imagination may be creating the conditions for the crystallization of a more general insight.

Since Newton, this unfinished chain of ideas has been accompanied by corresponding changes in the type of mathematical calculus used to represent the ideas quantitatively so that they lead to the prediction of measurements. Newton used a geometrical representation of forces, Boscovich an atomic geometry (supplemented by some algebra), and others since various analytical calculi; e.g., differential equations, matrices, operators, groups, etc. These calculi, from Newton to Dirac, all depend ultimately on the use of linear coordinate systems, $(x, y, z; t)$ or $(x_1, x_2, x_3, x_4 = ict)$, to represent spatial and temporal relations, though in recent developments (since 1927) the basic laws also involve higher abstract manifolds.

This employment of increasingly abstract calculi has tended to draw the attention of theoreticians to special mathematical problems. But the physical aspects remain challenging. As we have seen, Einstein suggested in 1909 that e and h, both representing an atomicity or discreteness of interactions alien to Maxwell's theory of the electromagnetic field, should arise as elements in the derivation of electron theory and quantum theory from a generalization of Maxwell's theory capable of accounting for both forms of apparent discreteness. Thus the value of (e^2/hc) should appear as a condition of a unified derivation of electron theory and early quantum theory.

In spite of many attempts, the fifty years which have passed since 1909 have produced no solution to this problem. Indeed the task has grown more complex, for e may be inseparable from the mass spectrum of the known particles, and the finite lives of the unstable particles also form part of a great complex of fundamental atomic constants, linked by pure number ratios. The only genuine achievement since Einstein's suggestion, possibly leading towards a solution, is a clarification of attitude: these atomic properties are now recognized to be expressions of modes of *interaction* (two-body, and perhaps many-body). But the exploitation of this recognition in a more powerful unified method lies ahead.

The long pause which has occurred in relation to this fundamental problem and others associated with it—fifty years since 1909, thirty years since 1929—suggests that some mental block, in the form of a deep-lying traditional assumption which has now become not only redundant but definitely obstructive, must be preventing a more rapid advance.

An example of the kind of issue which may be involved is suggested by the only fundamental advance made since 1929: the discovery in 1957–1958 of the failure of the parity principle in certain particle interactions. This shows that there is less invariance in particle processes than is postulated in the established theories, and this in turn may imply that asymmetrical relations (in the logical sense) are more important than had previously been thought. It is therefore conceivable that a more powerful theory of particles may have to employ asymmetrical relations of various kinds in its primary expressions: temporal, as one-way process; spatial, as chirality; and so on. Thus the obstructive element in current theory may conceivably be a persisting excessive reliance on symmetry and invariance, and inadequate exploitation of asymmetry and variance in the forms, perhaps, of contrasts between earlier and later or between left and right.

This view receives substantial support from the "PCT theorem" established in 1956, which may mark the culmination of the search for symmetry and invariances. For it suggests that the next major advance should simultaneously throw light on three primary issues: reflection symmetry/asymmetry; the geometrical significance of the two signs of electric charge and their equivalence/non-equivalence; and the reversibility/non-reversibility of the direction of basic processes. This 1956 theorem looks like the logical frontier of quantum mechanics beyond which there may lie a new post-quantum realm in which principles of invariance are not treated as fundamental, but derived from more general conceptions. Thus it is not surprising that atomic physics is now experiencing a fundamental uncertainty.

Only one thing is certain regarding the next major step: it will present a shock to minds trained to accept existing methods as absolute. Discoveries in physics fall into two classes: those that produce delighted surprise by confirming in some unexpected manner ideas that have already become familiar, and those that at first appear incredible or intellectually shocking to the majority of competent physicists because they refute principles still taken for granted.

It is interesting to note that in times of rapid advance these shocks are rather frequent, one or two occurring in each decade. Between 1895 and 1926 the current atomic orthodoxy received at least seven such uncomfortable blows:

1896	Radioactivity	Becquerel
1900	Discovery of quantum	Planck
1905	Discovery of photon	Einstein
1911	Small atomic nucleus	Rutherford
1913	Discrete orbits	Bohr
1923–1926	Wave particles	de Broglie, Schrödinger
1926	Indeterminacy, statistical interpretation	Born, Heisenberg

After the discovery of unstable particles, from 1932 onwards, the continual appearance of new types of elementary particles presented no great stir, for it had become clear that new phenomena were likely in the high-energy realm. Thus no basic adjustments were forced on theoretical physicists from 1930 until 1957, when the discovery of the failure of the parity principle for certain particle interactions began to compel a new orientation. It is interesting that this experimental proof of an intrinsic unique chirality in certain particles gave physical theory a shock of precisely the same geometrical-kinematic character as Oersted's paradoxical discovery that electric currents possess the property of deflecting a magnet in an asymmetrical manner, and also as the recent (1951–1957) discovery of non-symmetric stress tensors in crystals.

Around 1930–1940 it was widely believed that quantum mechanics (as developed in 1924–1929) could treat all atomic problems, subject to one doubtful exception: very low temperature properties. It was not at that time adequately realized that the effective scope of quantum mechanics might be subject to major restrictions:

A. The theory could only be applied where adequate approximative methods were available, and these depended on the smallness relative to unity of certain dimensionless ratios which could not be given any theoretical interpretation.

B. The theory had only been tested up to certain limits (smallness of size and largeness of energy of the systems). The canonical form of the theory has failed beyond these limits.

There are at present differences of view regarding the importance of the modifications which will be necessary to overcome these limitations. It may prove that some of the growing points which have been mentioned will provide a transition to a more powerful method.

CHAPTER 6

Postscript: On an Ideal Marriage of Continuity and Discontinuity

ATOMISM is not an arbitrary idea which man has imposed on his image of nature, nor is it (yet) a fully objective reflection of structural facts. It is a method of investigation which can be continually improved because it corresponds to features present both in objective nature and in the human mind.

This process of improvement occurs in two sharply contrasted ways: by short steps in the light of theory and experiment, and by long shots in the dark which illuminate a target, sometimes far ahead. At present the theory of atomism glows with points of light, but fundamentally it is still in the dark. No one knows where it is going. Minor steps will doubtless continue—until a new basic orientation emerges.

But one can at least formulate a standard of clarity, here and now. *What would be the ideal form for an atomic theory in the light of what is already known?* There have been times when such a question would have been absurd, as between 1924 and 1930 when fundamental theory was gaining fresh light every few months. Moreover, it is futile to construct ideal languages, theories, and calculi, without being guided by an intelligent interpretation of the facts.

But there are other times, as now, when there are plenty of facts, and a pause has given evidence of the need and opportunity for a new direction. In such periods it is useful to define an objective because, right or wrong, this may provoke fresh thought if the objective is happily linked to current problems.

Writing in 1878 of the state of thermodynamics after Clausius, Clerk Maxwell called it "a science with secure foundations, clear definitions, and distinct boundaries." The security, clarity, and distinctness evident to Maxwell did not last long, but his remark sets a standard. Can we define an ideal atomic theory which would appear to us today to possess those qualities? Even if we have to use terms which are vitiated by traditional errors, that should not deter the attempt. Since this essay treats atomism not as a closed chapter of clear answers but as a continuing puzzle, it is reasonable to suggest what might be regarded as a satisfactory solution.

All fundamental physical theories seek to unify phenomena in terms of the very small. An atomic theory employs discrete entities, or some combination of these with continuous functions. We may regard as an *ideal atomic theory* one which is ideal in its atomic aspects, a theory which makes the minimal assumptions of discreteness, using the simplest conceivable physical units, all equivalent and therefore not functions of their environment. All the non-equivalences of phenomena, the asymmetries, the differences of magnitude, and the variety must then arise from the necessary character of their changing spatial arrangements, under some universal law of process. Just as the mathematical theory of the crystal groups derives the possibility of 230 distinct types from the simplest of discrete assumptions *applied to the given three-dimensional character of physical space,* so an ideal atomic theory should derive all numerical relations from a simple law or system of laws applied to the various kinds of systems that can exist in three-dimensional space. We are unlikely in this century to "explain away" the overriding fact that physical space has three dimensions; the task of XXth-century physics is therefore to get the most out of this fact.

These thoughts lead to a fusion of discontinuous and continuous aspects which, I suggest, may appear the most natural to minds trained in our time:

Discrete point-centres displaying a continuum of changing spatial relations. Or, if we shift the emphasis from the points to their patterns: the clearest imaginable conception of variable structure is that of a continuum of changing spatial relations whose termini are discrete points. "Discontinuity" is then evident in the existence of discrete point-centres localizing and terminating the spatial relations, and "continuity" in the uninterrupted variation, in course of time, of these relations. This implies a basic distinction between spatial and temporal relations. The discreteness is a permanent micro-property in space and the continuity a spatially extended and *relatively* macro-property in both space and time.

Employing these assumptions, we make the intellectual experiment of defining an Ideal Atomic Theory (called "T") as a physical theory satisfying these conditions:

A. T must employ a single class of primary discrete point particles, all equivalent, possessing no quantitative properties.

B. All quantitative properties must arise from the three-dimensional spatial arrangements, constant or changing, of systems composed of a finite number of such primary particles.

C. A mathematical law representing these changing

point arrangements must provide a unified theory of all quantitative aspects of their order and disorder in systems of different types, including:

1. The known Quantum, Relativistic, and Classical Laws, as special cases present under particular conditions; and

2. The forms of order in stable complex systems and in systems in course of transformation.

The formulation of these conditions suggests the following queries:

What known experimental facts or current theoretical assumptions make T appear inconceivable or inappropriate? Can these features be reinterpreted so as to render less improbable an advance towards T?

Can the known particles be classified so as to build a bridge towards T? Can we, echoing Pythagoras and Kronecker, say, "God made the nucleons; all else is the work of man"?

Should the basic laws of particle interaction in T be:

Two-body or many-body?
Reversible or irreversible?
Scale-free or scale-fixed?
Dimensional or dimensionless?
Mechanical, kinematic, or geometrical?
Expressed in linear coordinates or in some other manner?
Employ scalar, polar, axial, or chiral magnitudes?
Invariants or variants?

Or do these antitheses miss some crucial point?

To these we may add a teasing question to which every exact scientist can provide his own answer:

In what sense, if at all, do the known phenomena *compel* an atomistic interpretation? Does T express this indispensable minimal atomism? If not, what is this necessary kernel of atomism?

One of the most cautious of living physicists, P. W. Bridgman, has said (1960): "Today the thesis of the atomic constitution of matter is universally accepted with no reservations whatever by every competent physicist."

This is manifestly true, in a certain sense. If we try to clarify its meaning and to discover what precisely it is that is thus universally accepted, we find, I suggest, that:

At a certain level of analysis, physical systems display a discrete or atomic constitution. This level is that of chemical molecules and atoms at normal energies.

But this atomic character may or may not be present at a deeper level, in smaller regions, or at higher energies. At the most fundamental and general level it is not certain whether physical systems are atomic, and in what precise sense. For a relativistic field theoretician, thinking along the lines laid down by Einstein and his followers, may hold that the primary phenomenon is a relational field of energy and that apparent discreteness is a secondary and transitory effect arising only in special circumstances.

The open question is whether the field or its sources is theoretically primary. It is already clear that at many levels discrete and effectively permanent structures are necessary; but are they at the level at which the next great theory must be built? After the long and successful history of atomism, it would certainly be surprising if the quantum mechanics of 1925–1928, with its subsequent variants, had exhausted the contribution which atomic ideas can make to fundamental theory. But this is a weak argument; we have seen that surprises are the daily bread of constructive periods in physics.

Atomic physics is spurred by the a-logical conviction that ultimate structure exists, is within human reach, and is simple (in the sense that it involves relatively few parameters). It is not necessarily true, as has sometimes been suggested, that no theory can ever be final and therefore no basic structure can ever be reached. Such structure may exist and may be identified, but long-sustained investigation will certainly be necessary before conclusions to that effect can be reliable and convincing. If some day a unified theory, resting on postulates regarding ultimate discrete structure, maintains its adequacy for a few decades in the face of all the facts including those discovered after its formulation, then it will be reasonable to assume that ultimate structure has actually been uncovered. It is the success of the laws built on a certain atomic structure which proves the ultimate character of that structure, and it is conceivable that a structure will one day be found which is ultimate in relation to all the known facts *and which maintains that status through the years and decades.*

But since 1850 no claim to finality made on behalf of any theory of fundamental structure has survived more than twenty or thirty years.

Annotated Selected Bibliography

THERE is an immense literature on atomism in several languages from around 1650 onwards. Apart from the more specialized treatments dealing with atomism in physics and chemistry, countless interesting passages on atomic ideas are scattered through works on philosophy, philosophy of science, history of ideas, and religion. A list of the most important discussions by the leading scientists and philosophers of the XIXth century alone might include a hundred references. No attempt will be made here to cover this vast field.

The following Selected Bibliography covers only the most valuable historical surveys of atomism in English, German, or French made during the last hundred years, i.e. since 1860, and a few books dealing with thinkers of particular importance for the history of the changing conceptions of atomism. Supplemented by the Chronological Table of the present work, this selection gives the historically oriented student of atomism the best available general coverage of the mainstream of European physical atomism up to the close of the XIXth century. Whittaker's *History of Theories of the Aether and Electricity* and volumes on Einstein and Bohr are included to provide a link with the particle physics of the XXth century, which is not covered in this bibliography. On the relations of Greek, Indian, and

Islamic atomic speculations, which also lie outside the scope of this selection, see S. Pines, *Beiträge zur Islamischen Atomenlehre*. Dissertation, Berlin, Gräfenhainichen, 1938.

As stated in Chapter I, there is not at present available in any language a history of atomic ideas written in this century and meeting the requirements of the teacher or student of physics or of the history of ideas. Gregory's *Short History of Atomism* (1926) perhaps comes nearest to this, but it is out of print and lacks the philosophical and mathematical precision which is desirable today in treating so powerful an idea.

I wish particularly to recommend to those interested in the history of scientific ideas or of atomism the three relatively short studies by Marie Boas, Partington, and Vavilov.

Brief notes on each work will assist the reader in choosing what he requires.

C. Bailey, *The Greek Atomists and Epicurus*. Oxford, 1928. A detailed survey of the Greek texts dealing with atomism, and of their authors, by a classical scholar.

M. Boas, "The Establishment of the Mechanical Philosophy." *Osiris, 10,* 412–541, 1952. This outstanding 130-page paper on the origins of the mechanical doctrine of atoms in motion deals mainly with Boyle and XVIIth-century atomism, but also outlines ancient and mediaeval ideas and includes references to the XVIIIth century. See also M. Boas on Hero, *Isis, 40,* 38–48, 1949.

G. T. Fechner, *Über die physikalische und philosophische Atomlehre*. 2nd edition, Leipzig, 1864. A remarkable work by the famous physicist and psychologist who greatly influenced Freud. It is not in form a historical survey, but it is included as evidence of the high level of philosophical and epistemological sophistication with which atomic ideas could be treated a hundred years ago.

J. C. Gregory, *A Short History of Atomism*. London, 1931. This attractively written semi-popular survey gives an interesting sketch both of the historical background and of the major steps in the development of physical and chemical atomism from the Greeks to the XIXth century. Should be supplemented by more recent scholarship (Partington, Boas). Out of print.

T. S. Kuhn, "Robert Boyle and Structural Chemistry in the Seventeenth Century." *Isis, 43,* 12–36, 1952. Contrasts particulate chemical theories with basic dynamical atomism. Useful for references.

F. A. Lange, *History of Materialism.* 3 vols in one (trans. 1879–1880 from 3rd German edition). New York, Humanities Press, 1950. The famous history of materialism and conceptions of matter from earliest times to mid-XIXth century. Useful, but more concerned with materialism as a philosophy than with changing atomic ideas.

K. Lasswitz, *Geschichte der Atomistik.* 2 vols, Hamburg, 1890. The most comprehensive survey of the history of atomism from the Middle Ages to Newton, with discussions of Greek, Moslem, and Indian ideas. Though some aspects which are of interest today are neglected, this work is a unique achievement and provides the basis for all subsequent historical research.

H. M. Leicester, *The Historical Background of Chemistry.* New York, John Wiley & Sons, 1956. Includes a useful survey of the development of atomic ideas in chemistry.

H. M. Mabilleau, *Histoire de la philosophie atomistique.* Paris, Imprimerie Nationale, 1895. An interesting interpretation of the earlier history of the idea of atomism, mainly covering Hindu, Greek, and mediaeval conceptions.

J. Clerk Maxwell, "Atom." *Encyclopaedia Britannica,* 9th edition, 1875 (reproduced under "Molecule" in 11th edition, 1910). This three-thousand-word historical sketch of the idea is one of the best available, as might be expected.

A. G. M. van Melsen, *From Atomos to Atom: The History of the Concept Atom.* Pittsburgh, Duquesne University Press. This interesting work, including original studies of Aristotle's *minima* and other matters lying somewhat outside the mainstream of scientific atomism, is valuable for students of the history of ideas. Less reliable on the philosophy of physics.

J. R. Partington, "The Origins of Atomic Theory." *Annals of Science, 4,* 245–282, 1939. The best concise survey known to me, from Democritus to Dalton, including some consideration of mediaeval, Moslem, Indian, and other schools, and a close analysis of Dalton's immediate predecessors.

W. Pauli, editor, *Niels Bohr and the Development of Physics.* New York, McGraw-Hill, 1955. A valuable symposium on the development of atomic and nuclear physics, mainly 1910 to 1950.

S. Sambursky, *The Physical World of the Greeks.* London, Routledge, 1956. Chapter V discusses the atomic ideas of Leucippus, Democritus, Epicurus, and Lucretius.

P. A. Schilpp, editor, *A. Einstein, Philosopher-Scientist (1879–1955).* New York, Harper, 2 vols, 1959. A unique symposium on the ideas connected with Einstein's contributions to basic theory.

G. B. Stones, "The Atomic View of Matter in the XVth, XVIth, and XVIIth Centuries." *Isis, 10,* 444–465, 1928. A useful survey from 1400 up to Boyle.

S. I. Vavilov, *Newton and the Atomic Theory.* Essay in Royal Society Newton Tercentenary Celebration volume, Cambridge University Press, 1947. An indispensable discussion, with quotations, of Newton's developing views and work on atomism.

E. T. Whittaker, *History of Theories of Aether and Electricity.* 2 vols: Vol I, *Classical Theories;* Vol II, *Modern Theories, 1900–1926.* London, Thomas Nelson & Sons; New York, Philosophical Library, 1951–1953. A valuable standard work, particularly on the mathematical aspects of electrical particles to 1926.

L. L. Whyte, editor, *R. J. Boscovich (1711–1787) Essays on His Life and Work*. London, Allen and Unwin, 1961. This symposium contains two essays on Boscovich's atomism, by the editor and by Z. Marković of Zagreb, including material on the history of atomism.